Identity

Identity

What DNA Can Tell Us About Ourselves

Carles Lalueza-Fox

The MIT Press
Cambridge, Massachusetts
London, England

The MIT Press
Massachusetts Institute of Technology
77 Massachusetts Avenue, Cambridge, MA 02139
mitpress.mit.edu

The MIT Press would like to thank the anonymous peer reviewers who provided comments on drafts of this book. The generous work of academic experts is essential for establishing the authority and quality of our publications. We acknowledge with gratitude the contributions of these otherwise uncredited readers.

This book was set in Stone Serif and Stone Sans by Westchester Publishing Services. Printed and bound in the United States of America.

Library of Congress Cataloging-in-Publication Data is available.

ISBN: 978-0-262-55322-3

10 9 8 7 6 5 4 3 2 1

EU Authorised Representative: Easy Access System Europe, Mustamäe tee 50, 10621 Tallinn, Estonia | Email: gpsr.requests@easproject.com

Contents

Preface

I am that I am.
—Exodus 3:14

How do I define myself? And how do I define others? Is there an objective way to do so? Simple as such questions may appear, anyone reading this book will run into trouble coming up with a solid answer. I can say I am a man, maybe a white, middle-aged man, a cultural Catholic, a son, a father; I'm European, Mediterranean, Spanish, Catalan; I am a scientist as well, and a reader, and perhaps even a writer. Am I English too? I had an English grandfather—who I never met, although I was named after him. Should I feel slightly English then (not an inconsequential question since under some circumstances, having a British grandparent may make one eligible for citizenship in the United Kingdom)? Nevertheless, if I try to give a definition for each of these items, I run into obvious difficulty. What does it mean, for instance, to be European? And what is it to be English? Is it simply something stated on one's passport or is it present in a fraction of one's genes? And which of these identities, if any, would best define *me*?

Saint Augustine (354–430) of Hippo had a famous dictum about time: "What then is time? Provided that nobody asks me, I know; but if I want to explain to an inquirer, I do not know." I have the feeling this could also be applied to one's own identity. If we were to actually examine the different definitions of identity, we'd soon descend into murky intellectual territory. And yet such concepts are of the utmost importance to us, and not just in adolescence when we attempt to construct an identity but repeatedly throughout our lives too, perhaps even up to our final days.

This book is in some ways a natural continuation of my previous one, *Inequality: A Genetic History*, where I explored how past inequality shapes our genomes.[1] In fact, where people are discriminated against in unequal societies, both as individuals or collectively, an obvious social mechanism in effect is the identification of some people as "others" and some as "ourselves." These classifications, whether based on physical or cultural traits, operated strongly in the past, and new, more subtle forms of identification have come into play in today's supposedly democratic societies that embrace diversity.

Some argue that following the 2019–2022 COVID-19 pandemic, we're facing a kind of end—or more precisely, a reformulation—of globalization (but clearly not a deglobalization).[2] In a world of imminent regional and global tensions along with critical scarcity of such basic resources as water, the issue of identity is bound to arise again as a crucial social landmark. In fact, we've already seen it unfold in the Russia-Ukraine war. On February 21, 2022, in a rather bizarre speech, Russian president Vladimir Putin argued against the legitimacy of current Ukrainian identity, claiming a deep, historical unity—as well as a common destiny—among the East Slavs (Russians, Ukrainians, and Belarussians).[3]

Because common heritage implies shared genetic ancestry—as well as shared cultural background—such a claim should be now scientifically testable. We are now able to retrieve thousands of ancient genomes from people who lived in the past—people contemporaneous to historical developments that shaped the modern world and could objectively help us to establish the genetic roots of collective identities. In 2023, a study estimated the number of ancient genomes to have surpassed the threshold of ten thousand, and they continue to grow exponentially.[4] These are remarkable numbers considering the first three were published in 2010. Obviously access to ancient genomes might influence certain notions of identity, including sociocultural ones rooted in the archaeological context.

The past overshadows our lives; it is long gone, but strangely solid in terms of identity. I was recently involved in a study that unraveled the genomic formation of the Balkans, a region, like Russia, Belarus, and Ukraine, where Slavic languages are spoken today and where some people identify themselves as Slavs—whatever that may actually mean. In this work, we retrieved genomes dating from the Roman Empire to the Middle Ages from such countries as Serbia, Croatia, Montenegro, North Macedonia, and Greece. We found that modern Serbians have a northeastern

European genetic component that arrived in the Balkans a thousand years ago; this arrival can be chronologically and archaeologically identified with the "Slavic migrations" that brought Slavic languages to the region. Yet the same ancestry poured into other regions of the Balkans in decreasing proportions as we move south. While this component constitutes around 40–60 percent of the modern population in Serbia along with neighboring countries such as Romania and Bulgaria, it might be 30–40 percent in Croatia, Montenegro, and North Macedonia, and only 20 percent in mainland Greece.[5] Why would such findings be taken as a sign of collective identity in Serbia but not in its neighbors? How do you establish a turning point in a gradient? And though Serbs and Croats may feel essentially different, it turns out they share a common genetic background. The most salient differences between the two communities seem to be cultural and religious (Orthodox church in Serbia and Catholic in Croatia), embodied in a long history of conflict. While ancestry and sociocultural notions of identity are often conflated, these results emphasize their independence.

When I presented our findings to the National Academy of Sciences in Belgrade in summer 2022, invited by my Serbian friends Miodrag Grbić, Illija Mikic, Dušan Keckarević, and Zeljko Tomanović, some academics in the audience voiced their disagreement with what they perceived as an undermining of Serbian "Slavism." Although I could not understand Serbo-Croatian, I could see they weren't at all pleased. In a country where 66 percent of Serbians called Russia the country's "greatest friend" in a poll at the beginning of 2023, the ambiguous and impartial attitude of the Serbian government toward the Russian invasion of Ukraine should come as no surprise.[6] When we published our results in late 2023, however, a Romanian academic wrote me blaming our study for promoting pro-Russian "Slavism" in this Latin-speaking country (unlike Serbians and Croatians, Romanians do not speak a Slavic language), and a British linguist criticized us for promoting an "ethnic" view of European history. I then understood the spurious aphorism attributed to Winston Churchill (1874–1965): "The Balkans produce more history than it can consume." And I realized genetics—particularly ancient genetics—had the potential not only to clarify mistaken preconceptions of human identity but also to provoke conflicting interpretations of "geneticized" identities that can have social and political implications. Is it fair that genetic researchers hold such potential importance? Even more, is it possible to avoid holding it?

Similar studies I've participated in over the past few years drive home the complexity of the identity issue, superficially trivial as it may appear, and it has begun to concern me. New genetic results have started to challenge our conceptions of identity in many different ways. Among other timely discoveries, the first gapless human genome was published in 2022, the first large ancient human pedigrees from the archaeological record in 2022 and 2023, and the first studies on genomic differences between twin and look-alike pairs in 2021 and 2022, respectively. The advent of large-scale computing algorithms has enabled us to not only explore the real genealogies of millions of people—in 2023, researchers published the genetic history of more than five million French Canadians over four hundred years—but understand the randomness of genetic transmission in human genealogies as well.

At the same time, companies devoted to tracing ancestries have generated millions of genomes from private test customers worldwide in the past decade.[7] In the United States alone, it has been estimated that 21 percent of adults—more than fifty million people—had taken such tests by 2023.[8] The resulting information has often been interpreted as a new way to understand identity—in fact, "true" identity. Moreover, these companies have used these genetic data to identify millions of distant relatives of their customers, effectively connecting people across the globe beyond standard genealogical registers. Yet as personal genetic data accumulate, we've seen the consequences of such enormous quantities of information—for instance, in the identification of criminal suspects through the search of genetic profiles of distant relatives. Coupled with new developments in forensic techniques that have enabled the retrieval for the first time in 2023 of whole genomes from such historical figures as composer Ludwig van Beethoven, it appears we now possess an unprecedented genetic knowledge of past and present-day individuals and populations.

In many cases, genetics has also challenged our assumptions about past human identities, such as the discovery that an important Viking warrior buried with abundant weaponry was in fact a woman; the finding that a pair of skeletons buried hand in hand and named the "Lovers of Modena" were both male; the unmasking of false relics from the French Bourbons; and the revelation that extrapair paternities occurred even in European royal lineages. Nevertheless, we're still far from agreeing over the scientific, social, legal, and ethical implications of all of this information, not to mention what to make of it in terms of our personal and collective identities.

Some of the ideas deriving from these works are fascinating. To mention just a few:

- Soon it will be possible to identify everyone from genetic databases, even those not in them

- Identical twins are likely not genetically identical

- Look-alikes, or doppelgängers, not only share certain genes associated with face morphology but also genes affecting behavioral traits

- The probability that we inherit any genomic fraction from a specific genealogical ancestor is almost zero after just some generations

- The number of our genetic ancestors is just a small subset of our genealogical ancestors

- In the past, many royals likely displayed genetic signs of endogamy, and in many cases their ancestry did not match the general ancestry of their subjects

- It is now possible to identify ancestors' relatives in the archaeological record, effectively linking people from the past to us by common lines of descent

- There are no "pure" populations; human genomes are always a composite of different, overlapping ancestry layers

- We are all interconnected through shared genetic ancestors in a gigantic and relatively recent network

Yielded from data that were impossible to obtain just fifteen years ago, such observations have the power to change our very notions of personal and collective identity.

This book, then, is an attempt to clarify what genomics can—and cannot—tell us about different levels of human identity. Genetic data, it must be kept in mind, are sensitive information that can be misleadingly used, such as by nationalist populist movements. There was a period after the completion of the Human Genome Project when some hoped genomics would clarify a number of debates about human nature. This notion, partly supported by an appeal to the authoritative nature of genetics, turned out to be optimistic, perhaps because the answers genetics provides about our identity sometimes do not match the questions we ask.

Still, genetics has the power to reveal our ultimate level of individuality—a level that makes each of us unique in human history. Genetics can reconstruct large genealogies, connecting contemporaries beyond the known family record. In fact, it has the potential to detect relatives that share common ancestors hundreds or even thousands of years ago, effectively bringing the past into the present—even if your chromosomal fragments are of Neanderthal origin. At the level of collective identity, although genetics cannot be used to assert the existence of races or ethnic groups, it is a tool capable of distinguishing the different layers of ancestry that comprise our genome, like a patchwork from the past. Moreover, genetics can in part be used to identify individuals as members of populations, even though those populations may vary in how well they are differentiated or defined. All of which has practical, social, and political implications. To take a couple of recent examples, a growing number of North Americans, as customers of genetic ancestry companies, are claiming to have First Peoples heritage, and recently, Irish citizenship was awarded to an adopted man born in the United States on the basis of a genetic test.[9]

In this book, I am intending to explore different levels of interaction between genetics and identity going from the individual level to families and genealogies, and from there to higher levels of collective identities (clans, populations, races, and even the human species). As we will see in these chapters, genetics—including ancient genetics—is a transformative science that challenges many conceptions, not only of identity, but of relatedness and ancestry. It is time to examine this new information and find out what our DNA has to tell us about ourselves.

1 The Entangled Nature of Our Multiple Identities

When I discover who I am, I'll be free.

—Ralph Ellison, *Invisible Man*

"Who am I?" is perhaps the definitive question in anyone's existence. It is asked countless times throughout our lives, no matter how old we are, whether standing before the mirror, facing a vital decision, or simply observing our surroundings. And though the very question seems to imply that a person can be just one thing, it likely has no single, solid answer. This is because it is rooted in our own personal identity, something multidimensional, complex, subtly evolving, and in many ways a mystery. And along with the question (which might be rephrased as "What is my identity?"), it is likely that other related questions arise that are as difficult as the first, such as "Who is like me?" or "Who do my peers say I am?"[1] The answer to the first question usually determines the others, and sometimes these reinforce the first in an identity loop. The importance we ascribe to such matters demonstrates the fundamental human need for identity.

Anyone can simultaneously have many different identities, which may be basic (that is, determined even before birth) or acquired during one's lifetime. Some of these identities may conflict or even contradict each other, but they define us as people and members of a collective. The contradictions in one's identity may well exist. As Walt Whitman (1819–1892) put it, "Do I contradict myself? / Very well then I contradict myself, / (I am large, I contain multitudes.)" At the individual level, there are numerous examples of flagrant contradictory identities: mafia members who condone assassinations with Catholic beliefs, white supremacists who are gay or marry Black people, and so on.

Identity can be defined at an individual level, concerning who we are, or think we are, but it is also inherently tied to a notion of social nature, wherein a person identifies with a particular human collective that stands apart from others. Identity, then, matters both for how we perceive ourselves and how others perceive us. This distinction is important because the former may not match the latter, whether as individuals or members of a collective. Identity is framed by both individual and population (or group) perspectives in a variety of contexts, and thus has crucial social, political, and cultural dimensions.[2]

Collective Identities

The diverse social constraints of identity have been expressed by Spanish philosopher José Ortega y Gasset (1883–1955) in his famous dictum "I am I and my circumstance." The notion of collective identity is directly related to the sense of belonging. If a person holds a particular identity that is shared by other people, they form a group to which they belong. Nothing else is needed but the mutual acknowledgment of belonging, subjective as it may be. This imagined unity of people can be built through a process that philosopher Charles Taylor (1931–) called the "politics of recognition" with regard to shared customs, traditions, and language, but also by dint of inhabiting the same geographic region. Ethnicity, race, and nation are examples of collective identities.

Once people ascribe or are assigned to a particular group identity, this automatically determines their values and worldview, and with them two important kinds of behavior: how to treat others and how to be treated by others. People who share identities are more likely to help one another. Collective identities, like individual ones, can have their own contradictions. For instance, Indigenous Christians from places in Mexico or Polynesia are able to perceive their pre-European contact communities as both peaceful, idealistic utopias and periods of heathen barbarism. Thus identity may be contradictory, and may alternately represent sameness and difference, imposition and choice. It may be singular or fractured, static or fluid.[3]

Group-level identity, though intangible and even contradictory, has its own appeal because belonging is often more about emotional attachment to a group than exclusively kin related. Such feelings of belonging are currently exploited in civil society, such as by sports clubs and political parties, as a mechanism for social cohesion.[4] The motivations and processes under-lying such identifications within groups are complex and multilayered, and include aspects like self-esteem and distinctiveness. In Nikolay Gogol's (1809–1852) *Taras Bulba*, the hetman (leader of a rogue band of Cossacks) and main character vehemently stresses the idea of belonging to a group over mere blood links: "Such was the time, comrades, when we joined hands in a brotherhood: that is what our fellowship consists in. There is no more sacred brotherhood. The father loves his children, the mother loves her children, the children love their father and mother; but this is not like that, brothers. The wild beast also loves its young. But a man can be related only by similarity of mind and not of blood. There have been brotherhoods in other lands, but never any such brotherhoods as on our Russian soil."

The feelings of belonging expressed by Taras Bulba find expression in historically produced nationalisms. Nationalism is concerned with death and immortality in a rather religious imagining, yet even if it focuses on a brilliant future, its roots are anchored in the past. As political scientist Benedict Anderson explains in his influential book *Imagined Communities*, "If nation-states are widely conceded to be 'new' and 'historical,' the nations to which they give political expression always loom out of an immemorial past."[5] Anderson argues that print media, along with institutions such as the census, maps, and museums, have formed modern conceptions of nationalism. Although the idea of imagined communities was applied by Anderson to nationalism, the concept is now routinely used to refer to other cultural and social communities, such as those united by race or gender. Anderson's work raises the point that sociocultural concepts of identity may also change in the future as technology evolves. And genetics, one of the leading scientific fields in our century, may well play a role in future transformations of identity.

In a highly influential work, Rogers Brubaker and Frederick Cooper maintain that the term "identity" now has too many meanings to be useful for sociological analysis, and instead propose to break it down into different categories.[6] They contend that "identity" can be categorized as both practical (defined by everyday social experience and developed by laypeople) and analytic (a scientific usage favored by sociologists). Aside from these categories deemed important by the authors, we can agree that identity has many other meanings within such contexts as social status, politics, culture, race, ethnicity, gender, sex, and nationality. And all of these identities are subject to a social process of reification—that is, treating abstractly defined concepts as if solidly existing in reality.

Identity can be actively sought out, and the search is often rooted in the past, which can be perceived as the foundation of the present itself—as well as the unknowable future.[7] This idea of the past as the substratum of our identity is reflected in genealogical research, which reaches as far back as biblical times with the tracing of priestly lines and royal dynasties.[8] Far from being an antiquated pastime, the field of genealogy now has a growing number of proponents across the globe who trace their antecedents through databases such as Ancestry.com and internet forums. People can then feel connected to the distant past as well as to their contemporaries who may be scattered across different continents. Genealogy is about seeking connections but, more important, it has an emotional component.

Thus identity may be linked to ancestry—that is, as far as heredity contributes to our existence, family, clan, or tribe, or even supraentities such as populations or nations. Such ambiguous perceptions remain oddly solid and long-lasting, and it is likely they've operated for millennia. Already in prehistoric times certain notions of complex kinship served to form ancestral social structures, such as the exchange networks among extended kin groups that functioned in sparsely populated areas.[9] Even relatively large genealogies might have been linked to a distant ancestor, real or mythical. Such a social mechanism has been described in New Guinean tribes as a way to avoid killing each other when two unrelated individuals met.[10] In modern societies, however, the demographic magnitude of human populations and existing cultural mechanisms are radically different, and are capable of forming identity on an unprecedented social scale. Fortunately, there is no need for killing each other, yet when two citizens from the same country meet in a foreign one, it is likely they will start sharing where they come from and even family details.

Social Sciences and Identity

The significance of identity in human communities has not only been investigated by social scientists. Philosophers as well have explored the interrelated meanings of human nature and identity. Opposing concepts linked to permanence and change, unity and diversity, are key to the philosophical layout of identity.[11] The Platonic school of philosophy saw humans as a duality, a combination of body and soul (with the latter having the search for knowledge and truth as its main task). But they also emphasized the social aspect of human nature. Christianity subsequently reformulated certain ideas of Plato (Greek philosophy was highly influential among early Christians), conceiving humans as possessing an immortal soul that is somehow trapped inside a body. Souls only existed in relation to God, who, according to Genesis 1:27, created them in his own image and likeness ("So God created man in His own image; in the image of God He created him; male and female He created them"). Aside from the Christian notion that human minds have nothing in common with animal minds since they were created separately, there is the idea that the design of women is based on that of men. (It must be kept in mind that in many Western countries, including the United States, this Abrahamic conception remains the most popular theory of human nature.)[12]

For philosopher and political theorist Karl Marx (1818–1883), comprehension of human nature was only possible through its imbrication in society, since humans are social beings. ("The human essence is no abstraction inherent in each individual. In its reality, it is the ensemble of social relations.")[13] For Marx, the idea of human individuality did not even exist. By contrast, the existentialists emphasized the individuality and freedom of humans, who have neither been created by any God nor have any purpose for their existence or any connection to another being. (According to Jean-Paul Sartre [1905–1980] in his book *Existentialism Is a Humanism*, "Man first of all exists, encounters himself, surges up in the world—and defines himself afterwards.")[14]

More recent debates about the meaning of human nature in the modern world seem to pivot on how we're integrated into the natural world and able to construct our own personality. One may argue that in recent times, the psychology of personality (described as the scientific study of the individual person) has in many ways replaced the intricacies of philosophy in the endeavor to define our identity. Many psychologists think of identity as a narrative based on one's life story that depends on the personality of the individual. Without exploring this level of personality, psychologists cannot determine how someone finds unity, purpose, or meaning in life. This life story, according to psychologist Dan P. McAdams, "is an internalized and evolving narrative of the self that incorporates the reconstructed past, perceived present, and anticipated future."[15] An interesting idea is that someone may have more than one life story, or a series of disconnected stories, about themselves. Whatever the case, a life story is a psychosocial construct—that is, it can be influenced by cultural environments as well as different temporal contexts—which means we might tell our story differently to our family than we would to our coworkers. For many personality psychologists, its function is to integrate the disparate (or even contradictory) elements at hand in order to produce a coherent and believable life narrative. It has two realms: one that accentuates our individuality and one that connects us to other groups of people. Therefore even from the psychological point of view, human identity has two different, recognizable dimensions: the individual and the collective.

In his 2002 book *The Blank Slate*, cognitive scientist Steven Pinker contends that human behavior and mind have been shaped by evolution, and the idea that we can be whatever we want, since our mind has no innate constraints (a condition termed a blank slate or tabula rasa), is fallacious and socially perilous.[16] Several cutting-edge scientific disciplines, Pinker claims, are taking the lead in building a new understanding of human nature that psychologists and philosophers alike should take into consideration. The science of mind integrates concepts from computer science and artificial intelligence. The cognitive neurosciences study the biological processes underlying cognition, and behavioral genetics explores the genes that influence human behaviors and their interaction with the environment. And evolutionary psychology contemplates human behavior as a set of psychological adaptations that can be examined on the basis of evolutionary biology. Of course, people might argue that knowing how the brain works when we take a decision or how our evolutionary history has shaped the way we grasp the world is not exactly the same as knowing our "true" identity, and thus the scientific study of mind is unlikely to be perceived as a path to understanding oneself.

Another major theme in Pinker's book is a rebuttal of Jean-Jacques Rousseau's (1712–1778) noble savage: the idea that society corrupts humans who are born free and naturally good. Also, Pinker questions another assumption about human nature that initially comes from René Descartes (1596–1650)—although clearly influenced by Plato—which he calls "the ghost in the machine." This is based on the idea that body and mind are different things and can be separated. Some maintain that the noble savage has made a comeback for the purposes of diversity, equity, and inclusion, and the ghost in the machine is widely used to support gender identity ideology.[17] Debates over empiricism versus rationalism in human nature go far beyond the scope of this book. The fact that they are still raging, however, shows the durability of notions about human nature and the resistance to changing them, or even, I would add, investigating them. But as Pinker points out, it goes beyond academic debate, since one conception or another can have important consequences for public social policies—including education.

Whatever the outcome of the seemingly endless nature-nurture debate—and I suspect most scholars would agree that the truth lies somewhere between the most extreme positions—we can agree that knowing a new dimension of human nature, the genetic level, will contribute to our understanding of identity, both as individuals and as populations, and also as a species. This does not mean supporting notions of identity that could be more "natural" or "scientific" than others but rather knowing the strengths and limitations of the genetic data.

The Arrival of a New Tool: Genetics

And as we have seen, different notions of identity, as explored by sociology, philosophy, and the behavioral sciences, always have a temporal dimension. In the twenty-first century, there are new, powerful ways to study human nature that are backed up by an unprecedented quantity of scientific data. Since the initial draft completion of the Human Genome Project in 2001, geneticists have been looking at the meaning of individuality and population, and even the limits of our species. Surprisingly, their findings have been largely ignored in the social sciences and humanities. A crucial observation is that the intermingled dimensions of individuality and collective identity traditionally examined by social scientists can both be approached from the perspective of genetics. With the existence of tens of millions of variable positions in the human genome, it can be argued that genetics holds the key to both the ultimate level of individuality and the limit to a species' diversity. And as genes help determine most aspects of our phenotype, there is also a complex interrelation between genetics and appearance (and then of course, how others see us).

In the last two decades, a new idea has emerged, both at the popular level and to some extent within academia—for instance, in the emphasis on the importance of personalized medicine—that an objective way of knowing oneself is to have one's genome analyzed by one or more of the many companies throughout the world that offer this service. By doing so, you might learn, among other things, where you fit in among the many different populations and ancestries of the human species. Inscribed in the DNA that packs your cells, this sort of information has been deemed the ultimate knowledge of life. And how better to know yourself than by looking deep within?

Our genome is structured in twenty-three pairs of chromosomes (one set from the mother and another from the father) and found inside the nucleus of each cell in our body (giving rise to the term "nuclear DNA"). In addition, multiple copies of a small circular DNA molecule are found outside the nucleus in cellular organelles called mitochondria (hence the name "mitochondrial DNA"), which are transmitted to offspring solely by the mother. The basic building blocks of DNA are the four nucleotides, each of which consists of a sugar molecule, phosphate group, and nitrogenous base: adenine (A), cyosine (C), guanine (G), and thymine (T). We have more than three billion nucleotides in our genome, which contains about twenty-five thousand genes, most of which code for specific proteins that function in one or more types of cells in our body. Our bodies are, in many ways, the outcomes of what our proteins do, from building tissues and body structures to acting as enzymes that control and participate in essentially all the body's chemical reactions.

Naturally, our DNA is inherited from our ancestors. Each generation buries the previous one, but our DNA goes on, scattered and fragmented, from the past to the future. Our genome is like a history book that conveys information that can be read on different timescales. We carry half the genome of each parent as well as DNA fragments from distant relatives—the longer the DNA fragments, the closer these relatives are. Many DNA sequences are identical to those of the species most closely related to our own, the chimpanzee, and even to more distant relatives. (Hence biochemist Jacques Monod's (1910–1976) aphorism, "Every living being is also a fossil.") Thus our DNA represents not only a promise for the generations to come but a complex heritage of past evolutionary events that can be discerned with the right analytic tools.

Genomics has progressed in several directions during the last few years: geneticists have sequenced hundreds of thousands of genomes from living individuals and stored their genetic information in large databases. In parallel, the advent of new sequencing technologies has allowed the routine retrieval of literally thousands of ancient human genomes from other periods—usually within the past few millennia—and a variety of geographic regions. This new scientific field is called "paleogenomics," or "ancient genomics." Genomes of people from the past help us to understand the DNA ancestry components of our own genomes as well as the spread of ancient migrations to different continents. At GoogleTrends, the term "DNA Ancestry" registered close to zero until 2013, when it began a continuous rise until peaking in 2017. Interest in the term has been slightly decreasing ever since (maybe because alternative concepts have taken its place), but nevertheless remains popular in Google searches.

Ancestry based on DNA has replaced ancestry based on "family links." The amount of data generated from DNA analysis is enormous, however, and needs to be interpreted. To summarize this entire body of information and capitalize on the growing interest in personal history, a new business model has recently developed: genetic ancestry test companies. Tens of millions have had their DNA tested, and a look at the distribution of customers worldwide makes it apparent that English-speaking countries are at the forefront of the trend; the United States leads the list, followed by New Zealand, Australia, Canada, and the United Kingdom.

Ancestry test companies play up the notion that learning about your genetic ancestry will help you understand who you are as a person and where you belong. After taking a small biological sample—usually of one's saliva—they return a report based on your genome that covers your ethnicity, physical traits, and the origins of your ancestors and even distant relatives among other customers from their own databases. They use ancient genomes published in recent genomic studies to include ancestry components that are labeled as "Viking" or "Roman." Such information offers customers a vivid picture that is in some ways more comprehensible and entertaining than what is provided by an actual genealogical profile, which in most cases goes no further back than a few hundred years of written records.

Uniparental Genetic Markers: The Maternal and Paternal Tales

Initially, some of these ancestry companies used uniparental markers such as the maternally inherited mitochondrial DNA and paternally inherited

Y chromosome (the latter carried only by males). At the dawn of this century, genome sequencing had to rely on painstakingly laborious techniques, and short mitochondrial DNA and specific markers on the Y chromosome were all that could be offered at a reasonable price. This, combined with the simplicity in their mode of inheritance, gave such markers some popularity until a decade ago. The Y chromosome might complement the genealogical surveys and surname research; the latter are so popular that some companies would do them for customers. Some companies sold this information with the idea of reconstructing customers' maternal or paternal line over time, simultaneously overemphasizing the importance of these two genetic markers. But they were not alone in this bias. Oxford geneticist Brian Sykes famously wrote the book *The Seven Daughters of Eve* about the seven main European lineages of mitochondrial DNA, fictionalizing the women who originally carried each of them.[18] By the way, the number of lineages is absolutely arbitrary as this comes from a genetic tree that could be "sliced" at any given level of branches, being two, seven, or thirty-seven. In fact, if we want to deal with the "main" branches of the human mitochondrial tree and fantasize about their founding women, we would need to be mostly restricted to the African continent.

As we trace our ancestors back in time, it must be kept in mind that both the mitochondrial DNA and Y chromosome markers come from just one ancestor each generation, while the potential number of ancestors that account for the rest of our genome doubles each generation (two parents, four grandparents, eight great-grandparents, etc). This means that seven generations ago, we had 128 positions in our ancestral genealogy, most of which were occupied by distinct ancestors, but our mitochondrial genome or Y chromosome will always derive from only one of these ancestors. Thus the proportion of ancestry that comes from our mitochondrial or Y chromosome ancestor gets smaller and smaller the further back in time we go (1 of 2 a generation back, 1 of 4 in two generations, . . . up to 1 of 128 seven generations ago). These uniparental markers can only represent the "maternal" or "paternal" genealogical line in a trivial sense, and their importance is sometimes overstated. In fact, the single ancestor they represent a given generation back—that is, the person carrying either our mitochondrial genome or Y chromosome—is no more significant for the resulting nuclear genome than the rest of the ancestors unrepresented by these two markers.[19] Moreover, if two individuals' uniparental markers match, the conclusion that this says something about their shared biogeographic ancestry can be problematic due to recent migrations. Without further genetic or historical information, the only sound conclusion is that the two individuals share a common ancestor in the past.[20] Interestingly, as the Y chromosome is about 57,000,000 nucleotides long and the mitochondrial genome only about 16,500, the temporal estimates for a common ancestor are different for the two markers. A complete identity in sequence of the Y chromosome in two males means they are closely related, while a complete identity in a mitochondrial sequence can place two related people within one thousand years of matrilineal relatives.

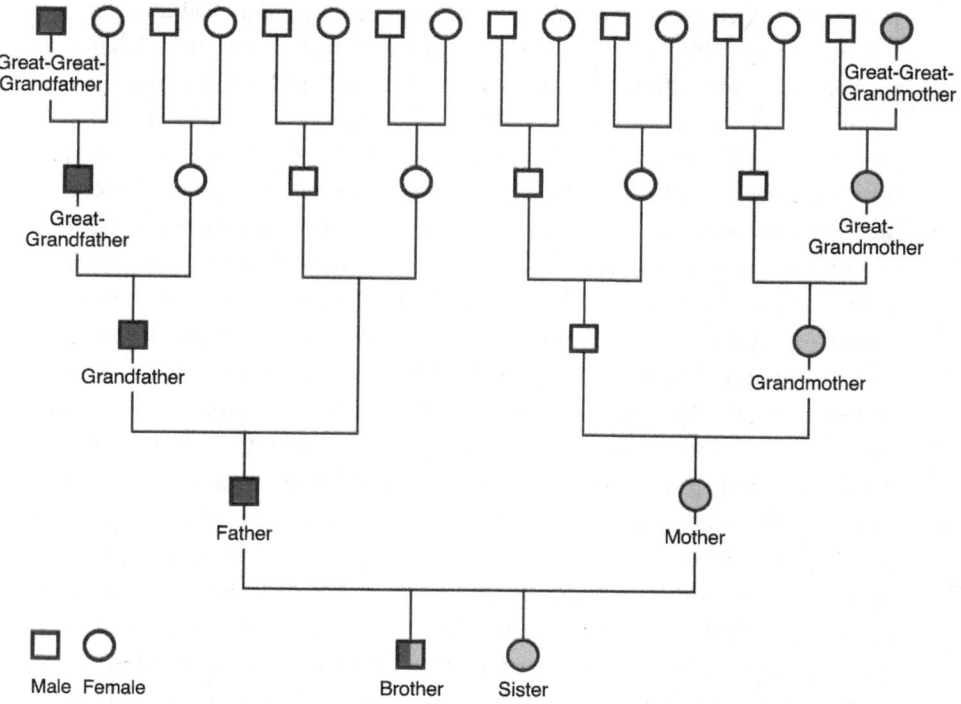

Figure 1.1

Differences between the inheritance of the uniparental markers (Y chromosome and mitochondrial DNA) versus the rest of the genome. While there are sixteen genealogical ancestors at the generation of our great-great-grandparents that have contributed to our nuclear genome, the Y chromosome (in black) derives from only one ancestor (the paternal great-great-grandfather), while our mitochondrial DNA (in gray) derives from only another ancestor (our maternal great-great-grandmother).

The misconceptions associated with uniparental markers have persisted beyond the arrival of ancient genomics. In a 1997 TV broadcast in Great Britain, Sykes, with the limited technical resources available at that time, attempted the retrieval of a small stretch of mitochondrial DNA from Britain's oldest near-complete human skeleton, the Cheddar Man from Gough's Cave in Somerset, dated to the mid- to late ninth millennium BCE. The results were never published in a peer-reviewed scientific journal. Afterward, Sykes sequenced the mitochondrial DNA of volunteers from the neighborhood and found a high school teacher born in Bristol, Adrian Targett, whose mitochondrial DNA matched Cheddar Man's.[21] It is indeed strange that such a misleading result made it to the news. Nine thousand years ago, Targett would have literally thousands of genealogical ancestors (we will explore this aspect in detail in chapter 6), and his presumed maternal connection would mean little in terms of shared genomic heritage.

More than twenty years later, with the new sequencing technologies, the whole genome of Cheddar Man was finally retrieved.[22] Like other Mesolithic genomes previously sequenced, the analysis revealed that Cheddar Man had blue eyes but dark skin—something Sykes would not have guessed twenty-five years ago. Cheddar Man bore scant resemblance to the local chap you'd find in a Somerset pub, and the depictions of Cheddar Man as a white, typically British individual at the Cheddar Museum of Prehistory sorely need updating. Amazingly, the locals, including Targett himself, now retired, were not disturbed by these new results. As one stated, "It doesn't matter what he looked like. What's color got to do with it?"[23] If anything, the story demonstrates the lasting impact that the news regarding human identity can have on the public, and how carefully preliminary results of genetic studies need to be explained. Indeed, the whole premise of Sykes's book as well as its development was misleading. But such dubious conclusions aren't limited to popular science books. As recently as 2019, the prestigious journal *Nature* published a paper that reconstructed and dated the most basal human mitochondrial DNA lineages—in South Africa, perhaps around the Okavango Delta—in which the authors claimed that our species originated in that area and period.[24] That some geneticists were still confusing the origin of a single genetic marker with that of the whole species at such a recent date is nothing short of astonishing. Yet it gives an idea of how powerful the mental image of a diversifying, branching tree can be as a temporal symbol offered by these markers.

A Look into the Past: The Ancestry Companies

Until recently, without genetic information from past human populations, it was difficult to deduce all of our ancestral links using contemporary DNA alone, despite innovative attempts such as those made by Luca Cavalli-Sforza (1922–2018) and colleagues. Human populations have been moving around for thousands of years, interacting with each other, overlapping, mixing, and sometimes replacing previous ones. Disentangling these long and complex processes has been the subject of the ancient genomics revolution. One immediate consequence of these studies is the acknowledgment that the current geographic distribution of uniparental genetic markers is not a good proxy for the location of that marker at any time in the past. One might suspect that a marker present in high frequency in a given population might be very old there. For instance, the fact that a particular Y chromosome lineage, the R1b, is the commonest in western Europe and highly prevalent in relatively isolated populations such as Basques made people believe it was the ancestral paternal lineage of the continent, perhaps deriving from Mesolithic hunter-gatherers. After all, up until a few years ago, Basques were thought of as a Paleolithic relict population. We now know that the R1b lineage arrived in eastern Europe with the Yamnaya steppe nomads about five thousand years ago and on the Iberian Peninsula no earlier than five hundred years later.[25] It is likely that this migration, coupled with the implementation of hierarchical social structures, accounts for the high prevalence of the lineage today and the disappearance of previous local lineages. The interpretation of the R1b distribution, in fact a relatively recent arrival into the European gene pool, would have been extremely difficult without the retrieval of literally thousands of ancient genomes.

At the same time, genomes from past populations offer the possibility to explore the origin of the components of genetic ancestry we observe in modern populations. Many people in Western countries now think their "true" identity can only be ascertained through their DNA. The special status of DNA as the final arbiter of truth of identity is reflected in the language used to describe it: "the code of codes," "the secret of life," the "instruction book," and so on.[26] And yet there are also conceptual and scientific limitations to the genetic approach to identity that need to be considered, opening new avenues for misconceptions over identity; it has been suggested that the genetic search for identity is related to a postmodern emphasis on radical individual self-determination.[27] Moreover, the genetic information needs to be interpreted within the broader context of population genetics, and this is not always easy to do, even for trained scientists, as we will see in chapter 7. The selection of categories or populations to which each customer is related is also arbitrary and often misleading, even if chosen in strictly geographic terms. Returning to Gogol, we could ask ourselves whether Cossacks are a genetically distinct population, as some Russian scientific publications suggest.[28] Or is that just a romanticized view of the seminomadic, rogue bandits who wandered the Pontic-Caspian steppe and no longer exist? We might ask too, What does it mean to be a Cossack nowadays? Is it something other than merely a historical construct?

Some years ago, I had a genetic analysis performed by an ancestry company that currently counts more than twelve million customers worldwide. The results showed I was 87.3 percent southern European, which broke down to 87.1 percent Spanish and Portuguese and 0.2 percent "broadly southern European." The rest of my genome was 12.7 percent northwestern European—which in turn breaks down to 6.2 percent "French and German" (a rather surprising category), 4.4 percent "British and Irish," and 2.1 percent "broadly northwestern European." Now considering some of my ancestors are Catalan (from Catalonia in the northeast corner of the Iberian Peninsula), and that population movements were prevalent in the past between Catalonia and southern France—the Roussillon was annexed by France as late as 1659—could it be that I share genomic sections with certain customers from that region? In fact, the family name of my maternal grandmother, Calvet, is native to the French Languedoc-Roussillon region. The selection by the company of a "French and German" category would in this case give a false impression of more "northerly" ancestry. Of course, the categories chosen by a company ultimately depend on the number and origin of its customers, and it rarely reveals its criteria. Even among the ancestry companies, there are quite large differences in the quality of results. For instance, in a recent blind test reported in the press, two identical twins took one of these ancestry tests, and while one was reported to be 13 percent "broadly European," the other yielded only 3 percent of the same category.[29] I assume this kind of discrepancy would be impossible in the many companies that have huge customer databases. It is worth emphasizing that solid scientific studies should yield more reliable conclusions than most ancestry companies.

The companies can indeed be quite accurate at identifying individuals among their databases who are related, likely up to the level of distant cousins.[30] According to my ancestry results, I have more than four hundred relatives in the United States, most of them at the level of fifth cousins (including three people from Hawaii, I was pleased to discover). But what this actually means is that I share, on average, a tiny fraction of 0.05 percent of the genome with each of them, which translates into a single DNA fragment about 12 million nucleotides long. Is this really trustworthy? I am quite confident of the reality of, say, second- to third-degree relatives, but finding distant relatives like those four hundred poses a technical problem that depends on the quality of the data produced.

The ancestry companies are also able to identify the geographic regions where their customers' ancestors lived with some precision. As we will see in chapter 7, however, there is by no means any certainty that their genealogical ancestors lived in regions where many of their descendants are found today. The availability of ancient genomes to compare with customer data means some companies try to make the link between present and past civilizations. They can break your DNA down into fractions labeled "Celts," "Romans," "Illyrians," or "Hittites." How the personal genetic composition is presented to the public is typically misleading and, being such a sensitive subject, can have consequences for perceptions of personal identity. In the past few years, I've been approached by tens of customers from these companies wanting to know what exactly is meant by their being "10 percent Minoan" or "5 percent Roman." These companies don't usually give access to the algorithms they use to estimate ancestry components and therefore population geneticists are unable to independently test their results, as they would in academic science.

With the availability of ancient hominin genomes such as those of the Neanderthals (the term "hominins" refers to the extinct members of the human lineage), such results also provide a Neanderthal ancestry estimate, which can go from zero in sub-Saharan Africans to almost 3 percent in non-African humans. My figure is 2.7 percent—13 percent over the average customer's. Here, again, how this heritage is interpreted remains ambiguous. If Neanderthals are thought of as "primitive," people will be happy the fraction is so small. Yet the fact that sub-Saharan African groups harbor almost no signs of Neanderthal admixing—because it took place in the Near or Middle East, after the initial out-of-Africa migration—has prompted some white supremacist groups to see this fraction as something positive, associated with strength, endurance, and ancestral adaptation to European environments. Some of them have adopted Paleo diet habits and are fond of such activities as weightlifting, which they identify with their Neanderthal ancestors.[31] The reality is that not only were Neanderthals not "primitive" in any sense (they had, for example, art, body ornaments, and funerary practices), but most of the genes we've inherited from them are also associated with neither strength nor diet.

The interpretation of genetic information provided by ancestry testing companies goes beyond the personal dimension; it could well usher in a new dimension of racial or collective identification. Race, especially in the United States, is increasingly perceived as an identity with multiple, sometimes conflicting dimensions that include, for instance, self-identification of racial identity, the race others believe you to be, or even the racial phenotypic appearance.[32] But test takers don't simply accept the results provided by companies at face value, taking them as a new form of "geneticized" identity. Instead, they selectively accept some results while rejecting others, according to their identity preferences as well as their social context. Interestingly, preexisting racial identities also influence the acceptance of the test's ancestry results. In the United States, self-reported whites are the most likely to accept "geneticized" identities, while those self-identified solely as Black are the least likely. And it seems Asians are less curious about their ancestry, perhaps because they perceive themselves as not racially mixed. Interestingly, customers only embrace those genetic identities that offer distinctiveness as well as social or psychological value.[33] Among white nationalists, there are a growing number of strategies to reinterpret test evidence of non-European ancestry and restore white identity among their members.[34] The conclusion is that despite the promise of scientifically objective truth from genomics, the test takers pick and choose the evidence they want from their results.

The Political and Social Implications of DNA

Playing with genetic ancestry to shape notions of collective identity is never an easy—or objective—task and can have significant consequences in contemporary politics.[35] A sensitive example arose in a 2019 genomic study, in which ten Iron Age skeletons from the ancient port of Ashkelon, in what is now southern Israel, were identified as "Philistine."[36] The researchers found that the Iron Age samples had a significant European-related genetic component that was absent in previous Late Bronze Age individuals, and they wrote in the paper, "This timing is in accord with estimates of the Philistines' arrival to the coast of the Levant, based on archaeological and textual records." Four days after publication, on July 7, 2019, Israeli prime minister Benjamin Netanyahu came down against Palestinians who apparently might have used the genetic results to claim proprietary right to the land. On Twitter, the prime minister wrote, "A new study of DNA recovered from an ancient Philistine site in the Israeli city of Ashkelon confirms what we know from the Bible—that the origin of the Philistines is in southern Europe." And he added, "The Bible mentions a place called Caphtor, which is probably modern-day Crete. There's no connection between the ancient Philistines and the modern Palestinians, whose ancestors came from the Arabian Peninsula to the Land of Israel thousands of years later."[37] His declarations precipitated a public debate with media uproar and accusations, and also the intervention of scientists. Harvard geneticist David Reich, who has a Jewish family background and was not involved in the original study, said, "Citation of genetic research in this way to support territorial claims is a truly sort of abhorrent and non-fact-based perspective of the past," and added, "It's really not clear who was there first. It's very clear from looking at the genetics that everybody in the Near East today, Palestinians, Jews have a lot of ancestry from people genetically similar to early Canaanites."[38] I think this episode demonstrates, like in the case of the Slavic migrations, how careful researchers have to be now with the presentation of genetic results in subjects with potential political and identity implications.

Another recent illustration of a complex admixture of cultural and genetic heritage that spurred a reevaluation of identity occurred in Mallorca, the largest of Spain's Balearic Islands. About six hundred years ago, there was a Jewish community on the island that numbered 1,435–4,000 members, among them the well-known rabbinical authority Simeon ben Zemah Duran (1361–1444). This Jewish community, which came to be known as Xuetes or Chuetas (the origin of the name is unclear, but it could come from "juetó," a diminutive for Jew, or, more implausibly, from "xulla," a Catalan word for a piece of bacon), was forcibly converted to Catholicism in 1435 and subsequently persecuted by the Spanish Inquisition over the next few decades. In 1691, several auto-da-fé condemned eighty-eight people, three of whom were burned at the stake.[39] The trials ended in the eighteenth century with the death penalty in 1720 of a Xueta who subsequently fled to Alexandria. Nevertheless, the institutional norm of prohibiting marriages outside the Xueta community, which comprised fifteen family names, persisted until recently. Despite their formal conversion to Catholicism, they were discriminated against and socially isolated until the mid-twentieth century.

In the past few decades, as the sense of community declined, there have been some attempts to recognize the Xuetes as Jewish. In 2011, for example, Rabbi Nissim Karelitz, head of the rabbinical court in Bnei Brak (Israel), officially ruled that "all descendants of converts from Mallorca who can prove that their maternal grandmother, before the Second World War, had as a last name of those 15 that belong to the Chuetas, must be considered Jewish."[40] (Karelitz's rule would ultimately not be upheld by the Israeli state rabbinate.) Paradoxically, now that the ostracism has ceased, the dangers of assimilation are greater than ever and, in fact, few of its members have officially converted back to Judaism, while even fewer went to live to Israel. Recent genetic studies on the Xuetes's descendants through the use of uniparental and autosomal forensic markers revealed Near Eastern affinities, but also signs of admixture with the local population, especially in the mitochondrial DNA lineages. (It must be kept in mind that Judaism is traditionally transmitted via one's mother). Interestingly, and despite hundreds of years of social isolation, the community does not show the genetic signs of endogamy usually found in small populations or endangered species.[41] Again, this suggests that some gene flow from outside the community took place.

What is interesting is that after six hundred years, some Xuetes's descendants felt their ancestral Jewish heritage was important enough to radically change a significant part of the cultural, social, and even geographic environment where their families and friends lived. For people growing up in the cosmopolitan environment of an international tourist destination like Mallorca, amid the growing secularism of Spanish youths, this is quite surprising. Of course, any search for identity is legitimate, and in the end any action constitutes a personal decision. If nothing else, this story speaks to the enormous psychological weight the past has on the present and the fluid perception of ancestry with regard to identity.

A similar scenario, although on a larger scale, relates to the African slave trade in the Americas. In the eighteenth to nineteenth century, some

European Americans sponsored a back-to-Africa movement in the belief that a significant number of African Americans forcibly transported there would choose to go back to the African continent—anywhere on it, apparently. The movement was a failure, and only a small number took the opportunity to emigrate to Liberia around 1820. Their mortality numbers were terrible. In a couple of decades, only 1,819 of the original 4,571 African American settlers were still alive.[42] In modern times, Black intellectuals such as Frank B. Wilderson III and Christina Sharpe have developed a line of thought known as "Afropessimism" by which not enough shared identity remains between the two communities of African Americans and Africans to create a basis for solidarity. As they see it, the struggle of the former group cannot be linked to the anticolonial struggles on the African continent. Alternatively, some prominent African Americans have tried to promote a kind of Pan-Africanism. In 2003, geneticist Rick Kittles launched a company named African Ancestry that claims to have helped more than a million Americans trace their roots to the African continent. Actor Isaiah Washington took an ancestry test that yielded a putative connection to the people of Sierra Leone; he subsequently gained dual citizenship based on this test, thus fulfilling what might be seen as a journey of personal reconciliation.[43]

Although some African countries have tried to attract African Americans, the numbers of emigrants from the United States to Africa remains low. (Ghana dubbed 2019 the "Year of Return" and since then has received around fifteen hundred migrants.)[44] Due to a novel combination of social and political factors, including the geopolitical shifting of African countries to China instead of the United States, and marked cultural differences between Africans and African Americans, Pan-Africanism is a movement that has largely faded away in the United States.[45] Furthermore, there remain certain methodological issues to consider. First, during the slave period, people of different origins were mixed, likely conforming a highly complex ancestral tapestry that cannot easily be attributed to a single "tribe" or geographic place. And even attempting to define current tribes is problematic, as is the concept itself. Some authors have attributed such designations to arrangements made by colonial powers, which split groups across artificial borders or grouped them according to such criteria as linguistic affiliations. The very supposition that African societies were structured into "tribes" that mirrored primitive European societies is likely wrong. Most African cultural groups were too small or too large to fit into the arbitrary ethnic divisions of the continent promoted by Europeans.[46] Also, Africa is both the most genetically diverse continent and the least studied. It is likely that trying to match, say, a million North American customers would require in turn a significantly larger number of African customers—perhaps ten times more. Regardless of how the attempt by Black Americans to reconnect to the African continent fares in the future, here is a case where genetic ancestry can have important social, economic, and even political implications.

To add further complexity to potential reparations for the enslavement suffered by the Black American community, the underlying genetic evidence is just now being explored. In the United States, several genetic studies have revealed that self-identified African Americans carry a significant degree of European ancestry—a quarter on average.[47] The European ancestry component increases toward the north. Overall, about 10 percent of self-reported African Americans have less than 50 percent African ancestry.[48] A small fraction (around 2 percent) of African Americans carry less than 2 percent of African ancestry and might be more appropriately described as white (according to the US census categories). A similar confusion of identity applies to former president Barack Obama; while a majority (55 percent) of African Americans see him as "Black"—and this is the option he checked on the US census—his ancestry indicates he could just as well be seen as "mixed race" or "white and black."[49] The European component was introduced into the Black American community about twelve generations ago, during the period when slavery was legal. A decrease of 5 percent in European ancestry in the X chromosome (women carry two X chromosomes and males only one) and the large presence of European Y chromosomes in Black American males indicate that this European component was male mediated. Some sources estimate that 58 percent of all enslaved African women between the ages of fifteen and thirty were raped by their owners or other white men.[50] The difference between the white and Black American communities is extreme; the researchers estimated that only 1.4 percent of self-reported European Americans have over 1 percent of African ancestry in their chromosomes.

Understandably, many African Americans feel uncomfortable about their white ancestry, when reported, as it likely derives from the history of sexual domination that has shaped their community. Certainly, this is a complex heritage to deal with. Equally troubling for them is the fact that many white Americans, including five living presidents—Obama among them— two Supreme Court justices, eleven governors, and a hundred legislators descend from slave owners.[51] Of course, one might well ask whether we have the right to hold distant ancestors to the moral norms of the present.

And how to deal with a more recent, evil heritage? Bettina Goering, the great-niece of prominent Nazi leader Hermann Goering (1897–1946), had herself sterilized at the age of thirty because "I feared I would create another monster."[52] In fact, being his great-niece, she would carry on average 12.5 percent of Goering's genome, and one might ask if this perceived "evil" fraction would dominate the remaining 87.5 percent. Such a low percentage may prove insignificant, although she thinks "it might be in my DNA. I think they're starting to prove that all experiences of your ancestors manifest themselves in the DNA."[53] (She might be referring to the epigenetic footprints that environmental variables can leave in one's DNA that are passed on to the next and subsequent generations. But the Nazi leader was not even her direct ancestor.)

If nothing else, these stories illustrate the importance we attribute to our ancestors in helping us find our place in the world. In many ways, the personal endurance of these symbols is astonishing. It relates to notions of psychological essentialism—that is, the belief that people have internal, immutable essences that influence their identity. The tendency to infer that such identity traits are based on our genetic makeup is called genetic essentialism, potentially a rising trend in the twenty-first century.[54]

All of these recent technical developments in genomics, along with a seemingly endless flow of information, are shaping notions of both individual and collective identity. In the era of postglobalization after the 2019–2022 COVID pandemic, human identity is emerging—once again—as an important way of seeing and organizing the world. And genetics, whether explained by scientific literature or interpreted by ancestry companies, will likely play a role in our future perception of identity.

2 Genetics of Individuality

The individual has always had to struggle to keep from being overwhelmed by the tribe.
—Rudyard Kipling, "Interview with an Immortal"

King Louis XVI was charged with treason by the National Convention on January 20, 1793, and guillotined the next morning on the Place de la Révolution (today Place de la Concorde, famous for having a Luxor obelisk installed there in 1836) before a crowd of twenty thousand. He spent his last afternoon with his wife, Marie Antoinette, his sister Élisabeth, and his children. He got up at 5 a.m. on a typically foggy, cold Parisian morning, and after getting dressed, passed out his few personal items (spectacles, ring, and royal seal) to the Irish priest Henry Essex Edgeworth and attended a brief mass. It took an hour and a half to go from the Temple prison to the scaffold. In the middle of the square, the equestrian statue of Louis XV had been toppled over, leaving an empty pedestal. The king might have seen this as a metaphor for the demise of his dynasty. Once there, the king, described by many courtiers as shy and irresolute, started a brief speech: "I die innocent of all the crimes laid to my charge; I pardon those who have occasioned my death; and I pray to God that the blood you are going to shed may never be visited by France."[1] Then the drums were ordered to beat and the king's words became inaudible. Moments later, he was dragged to the guillotine and quickly beheaded without further ceremony. The crowd rejoiced and burst into singing "La Marseillaise." Witnesses reported that many leaped to the scaffold to dip handkerchiefs and papers—and some even their fingers—in the king's blood. The execution of Louis XVI, which signified the end of the ancien régime, was a transformative and symbolic episode in European history.

The revolutionary authorities arranged for the body's disappearance. The corpse was brought to the Madeleine Cemetery and buried in an anonymous coffin, then covered with quicklime to ensure the destruction of the remains. Later that year, the ancient Royal Tombs at the Basilica of Saint-Denis, near Paris, were opened and all human remains dumped into a trench and covered with quicklime. With the Bourbon restoration in 1817, fragments of unidentifiable bodies were recovered and placed into two sealed ossuaries in the crypt, never to be opened again. Apparently, the bodily remains of the entire French dynasty had disappeared. Or hadn't they?

A French Royal Cold Case

Some years ago, I was involved in a genetic study of what might be called a *very* cold case. At the end of 2008, a friend of mine from the University of Bologna, Davide Pettener, contacted me with a rather unusual proposal. A rich and anonymous Italian family from Imola (Italy) held a precious artifact from the French Revolution and wanted to know more about it, particularly regarding its authenticity. It was a pyrographically decorated gourd that supposedly contained the blood of French king Louis XVI, guillotined in Paris on January 21, 1793. How did the blood of the king end up in a gourd in Italy? The gourd, which measured 23.7 centimeters long and 15.2 in diameter at the base, had the typical shape of *Cucurbita moschata*, sometimes used in traditional societies to carry water, but also in revolutionary times to hold gunpowder for soldiers armed with a rifle. It was decorated with numerous images and text. On several medallions, portraits of key royalist figures could be made out, including "Louis XVI roy des François," "Louis le Dauphin," "Necker," "M.[arie] A.[ntoinette] R.[eine] D.[es] F.[rançois]," "Simon," "Montseigneur" (?), and "F.[rère] D.[u] R.[oy]." Other medallions at the base portrayed such revolutionary figures as Georges Danton, Jean-Paul Marat, Camille Demoulins, Louis Sebastien Mercier, Joseph-Ignace Guillotin, Maximilien Robespierre, Marquis De Launay, Jacques de Flesselles, and Joseph Foullon. An inscription within gives crucial information about this intriguing object: "Maximilien Bourdaloue le 21 de Januier [*sic*] de cette année imbiba son mouchoir dans le sang de Louis XVI après sa decollation" (that is, this person claims to have dipped his handkerchief in the king's blood after he was publicly beheaded). In additional text, the owner further explains that he commissioned an artist to decorate the gourd and expresses his hope to sell it to "the Eagle" (possibly a young Napoléon who had just won notoriety at the siege of Toulon in 1793) for the amount of five hundred francs. There is still space enough to fill it with Masonic symbols as well as a rather enigmatic sentence that can be translated as, "Here lie those that didn't know how to live." It is difficult to track down the itinerary of the gourd from Paris to Imola. The family that owned the object also had a document from the Musée Carnavalet in Paris dating from 1900, when the Italian owners asked for a curator's opinion on the authenticity of the content that they claimed had been in their possession for about a hundred years.

All my friend Davide could see inside the gourd was a blackened substance but no obvious remains of a handkerchief. Using luminol, a solution commonly employed in forensics to detect blood by generating fluorescence, it could be confirmed that the dark matter was indeed very old blood. We then decided to use the most advanced ancient genetic techniques to ascertain if this blood did in fact belong to the French king. In several weeks of laboratory work, we were initially able to retrieve the mitochondrial DNA and a Y chromosome profile with a combination of sixteen variable genetic markers. The paternal chromosome was a rather uncommon G2a lineage. In any forensic identification work, however, it is essential to have someone to compare with, either a potential progenitor or potential suspect. But in this case, no such individual appeared since all the royal tombs at Saint Denis had been destroyed during the French Revolution. Although there are modern descendants of the Bourbon royal family—including, for instance, the king of Spain—I couldn't find any way to contact them for a genetic test; therefore I decided to publish the results, with the potential genetic profile of the French king, in a forensics journal, though without any real possibility for independent authentication.[2]

Yet things took an unexpected turn. Just a few years later, though I still had no living Bourbon to compare the remains with, I did have a potential ancestor: the mummified head of King Henry IV of France (1553–1610), nicknamed Good King Henry or Henry the Great (or le Verd Galant for his numerous love affairs). Henry was originally king of Navarre (as Henry III) before becoming king of France following the death of Henry III without an heir. He had a complicated life in a country devastated by religious wars.

During his wedding in Paris to Margaret of Valois, he barely escaped assassination in the St. Bartholomew's Day massacre (August 24, 1572), which saw the killing of thousands of Protestants who had come to the city to attend the ceremony. In 1593, Henry renounced Protestantism and converted to Catholicism—an act accompanied by a monument to political pragmatism with his statement (perhaps apocryphal), "Paris is well worth a Mass." In 1610, Henry was assassinated by a Catholic fanatic, François Ravaillac, and his body subsequently buried at Saint Denis. His remains, which turned out to be exceptionally well preserved and mummified, were exhumed in the sacking of the royal crypt during the French Revolution and publicly exposed before being anonymously buried in quicklime. Strangely, it seems the head of the king somehow avoided reburial—perhaps owing to Doctor Alexandre Lenoir, a witness to the desecration of Saint Denis—and emerged after some decades in the private collection of a German count, Franz Erbach, from Hesse-Darmstadt. In 1919, the head returned to France, where the photographer Joseph-Emile Bourdais acquired it for just three francs. On his death in 1947, the head vanished once again, only to resurface in 2008 in the house of an old couple, who it turned out had purchased it from Bourdais's daughter in 1955, this time for five thousand francs—a sign of inflation in the severed head's market, I guess.

A French forensics expert, Philippe Charlier, had the opportunity to examine the head and published a report positively identifying it as the king's—but otherwise providing negative genetic results. The radiocarbon analysis yielded dates between 1450 and 1650, and with computer tomography a likeness of the mummy was reconstructed, overlapped with a funerary mask of the king. Moreover, an old bone injury over the lip was compatible with a failed assassination attempt against Henry in 1594 (he was a beloved king, but there was, it seems, some interest in killing him). After the publication of this study, the owners of the head donated it to Luís Alfonso de Borbón Martínez Bordiu, a Spanish claimant to the French throne—one of many—who attempted unsuccessfully to have it buried at Saint Denis. In 2012, Charlier contacted me to obtain a genetic profile of the presumptive head of Henry IV, now that we presumably had the Y chromosome profile of one of his descendants, Louis XVI. If the two profiles matched, we could authenticate both the gourd and the head. After several weeks of work at my DNA laboratory, I was only able to obtain data from six markers out of the sixteen commonly used in forensics. Five matched the gourd's profile, and one was but a mutational step away, which was to be expected considering the seven generations that separated the two kings. The five markers had rather low frequencies in the modern European database, which reinforced the conclusion, even with the uncertainties of partial data, that the two individuals were related.[3]

This second identification collapsed like a house of cards, I'm afraid, when a subsequent genetic study of several living descendants from different Bourbon branches, conducted by Belgian forensics experts, demonstrated that Bourbon males had a genetic profile that was different from those I'd found. Furthermore, the researchers were able to prove that Marie Antoinette—wife of Louis XVI—and Henry IV must have had the same mitochondrial DNA since the latter was maternally related to the former through his mother, Jeanne III d'Albret. (Their common ancestor was Anne of Habsburg, who lived between 1280 and 1327!) The mitochondrial profile of Marie Antoinette was obtained from a historically authenticated lock of hair from the queen.[4] I was a bit frustrated for a while, but Christian relics were a flourishing market in the Middle Ages for reasons of economics and prestige, and these royal relics, though secular, fulfilled exactly the same needs. Our friend Bourdaloue, who clearly had a handkerchief with blood and wished to gain five hundred francs, could not have imagined what science would be able achieve in two hundred years. The blood could have been his own or possibly someone else's. After all, the guillotine never stopped in those years; in 1793, it took only thirty-six minutes to behead twenty-two Girondin members, which I guess illustrates the professionalization of the executers and the general availability of blood.

This story illustrates the dangerous limitations of forensic identification with the increasingly partial genetic profiles yielded by extremely degraded and technically challenging samples; for instance, potential twins and kin-related people—who will have similar genetic profiles to a suspect—can mislead the results. Not only can this be a problem with historical samples, it can, more worryingly, yield faulty results in real criminal cases when the technology is pushed to the limit. Also, a recent case of data manipulation during decades by DNA expert Yvonne "Missy" Woods has warned against the decision power exerted by some preeminent forensic scientists.[5]

Nevertheless, if genetic forensic data are properly generated and used, what accounts, we can ask ourselves, for the incredible discriminatory powers of current molecular techniques? It all stems from the level of variation existing in our genome, a feature used by forensics experts primarily to identify individuals. Forensics deals with the idea of individualization, but to understand individuals, we have to look into the human genome.

The Scale of Human Genetic Variation

The completion of the Human Genome Project was announced on April 14, 2003, but the information produced therein had some gaps all across the chromosomes due to repetitive regions that were impossible to read with the available sequencing technology. These regions, although not crucial from a medical or even evolutionary point of view, constituted around 8 percent of the genome. The final effort to produce a gapless human genome took place in 2022 with the publication of a complete 3.055-billion-base-pair genome, accounting for all chromosomes—with the exception of the Y chromosome, completed later that year.[6] The exact length of that particular genome was 3,054,815,472 nucleotides, or chemical building blocks, of DNA. Interestingly, the complete version of the genome showed the unmapped regions to contain 1,956 possible new genes, including 99 that were predicted to code for proteins. Although the genome constitutes a complete representation of human genetic information, it does not capture the whole range of human genetic variation. It has been estimated that any two random humans will differ in about 0.1 percent of their genomes, which accounts for one nucleotide in every 1,000 (or about 3,000,000 differences).

Considering their length, how much genetic difference can we reasonably expect to find in humanity's genomes? Obviously, the number of variable sites increases with the reference genome's sample size, so it's impossible to ascertain the precise figures with certainty. The ongoing 1000 Genomes Project (1KGP) uncovered the existence of 81 million single nucleotide variants, of which 68.4 million were rare variants found in less than 1 percent of the worldwide population.[7] Of the remaining variants, 2.7 million have frequencies of between 1 and 2 percent, 1.2 million between 2 and 3 percent, 760,000 between 3 and 4 percent, and so on (that is, the larger the frequency, the lower the number of variants). The subsequent stage of the same project (phase 3 of the 1KGP) increased the number of genomes and variable genetic variants to almost 92 million—91,784,637 to be precise.[8] Naturally, the higher the number of genomes sequenced, the higher the number of variable sites we're going to discover in the human genome. In the end, it is possible that all viable mutations not causing the death of the carrier will be present, including some recurrent ones that were present only in extinct hominin groups such as Neanderthals.

Of all genetic variants, those present at a frequency of at least 5 percent are called "common variants." These have been employed in a number of population applications, from identifying ancestries and relatives to revealing a genetic predisposition to common human diseases. In the extreme, this means that the existence of an individual genetic mutation—that is, only present in one single person—is of no value for interindividual comparisons or understanding of general population processes. Information gathered from the hundreds of genomic studies so far published suggests that about 10,000 genetic positions—mostly common variants—seem to be associated with the most common human phenotypes, including diseases.[9] Incidentally, the most common variants tend also to be the oldest in the human lineage because it takes time for mutations to spread across all continents.

Some studies have attempted to understand the world distribution of these common variants. Are they shared across continents or to some extent geographically restricted? One study, for instance, found 204,983 genetic variants roughly restricted to Africa, 46,994 to the Americas, 7,789 to East Asia, 6,585 to Europe, and 77,437 to Oceania.[10] This means that Africa is the most diverse continent and that the rest of the world, particularly Europe, is much less diverse. It is noteworthy, however, that nowhere in the genome does every individual native to a certain continent have a particular nucleotide that the rest of the world doesn't. The same can be said of any conceivable population or ethnic group, as we will see in chapter 7.

The Rise of DNA Forensics

One consequence of this enormous genetic diversity at the individual level is its application to identification. Although forensics is by now deeply intertwined with genetics and other technological advances, it wasn't until 1985 that Sir Alec Jeffreys (1950–), a molecular biologist from Leicester University, suggested that the concept of "genetic fingerprinting" be adopted in courts of law.[11] Since then, the discipline has made use of the incredible developments that have taken place in the field over the past two decades, including techniques associated with the retrieval of ancient DNA. Still, ethical and statistical issues remain substantial challenges, along with the aspects of how such data are interpreted by the public and how well they are understood when presented in court.

The current application of molecular approaches to identification is nothing new. At the beginning of the twentieth century, the discovery of polymorphisms in blood groups, starting with the ABO group, enabled some attempts at identification. They were mainly used to discard paternities, as some combinations of blood groups among parents were incompatible with others. (We can imagine, for instance, cases where both parents possess type O blood, indicating they can't have offspring of A, B, or AB blood types.) A famous court case was—wrongly—solved by means of this system in the 1940s, when Joan Barry claimed that Charles Chaplin was the father of her daughter, Carol Ann. To disprove her, Chaplin's lawyers asked all three to be typed for ABO. Joan had type A blood, her daughter type B, and Chaplin turned out to belong to the O blood group; thus he could not have been the father. Despite the incompatible results, Barry's attorney succeeded in arguing that the tests were inadmissible and Chaplin was obliged to support Carol Ann until her twenty-first birthday.[12]

The discovery of additional blood groups, such as MNS, Kell, and Kidd among others, provided additional discriminatory powers. The subsequent introduction of the highly variable human leucocyte antigen (HLA), which is key to organ transplant rejection, represented a boost in paternity tests after 1970. For years, the HLA system was screened by serological typing and therefore the underlying genetic variation could not be identified. Yet the biggest revolution in the field came with the discovery of DNA polymorphisms and the different techniques to screen them. The advent of the polymerase chain reaction technique (known as PCR), developed by Nobelist Kary Mullis in 1986, enabled forensic experts to work with increasingly small amounts of biological samples—such as blood, bodily fluids, hair, and so on. In a kind of competition between the presence of degraded DNA and discriminatory power, the field adopted the genotyping of short tandem repeats (STRs), which are markers comprised of a highly variable number of short repeating units (usually between two and seven nucleotides as the core repeated unit). As in the case of the blood groups, the combined screening of several of these variable STRs—up to twenty or thirty—greatly increased the specificity of each individual sample, to the point that no confusion was possible. Starting with the UK National DNA

Database in 1995, all technologically advanced countries established a DNA database of STR profiles, with which unknown profiles from crime scenes could be compared with known individuals, including, of course, suspects. In parallel, the International Society for Forensic Genetics standardized all experimental protocols and statistical methods—an effort that ensured the portability of the data from one country to another. The Federal Bureau of Investigation (FBI) has estimated the probability of two unrelated persons having a matching profile by means of the thirteen STRs commonly used in forensics to be on the order of one in a billion or even greater; in the case of seventeen STRs, the probability raises to about one in a trillion.[13]

The genetic markers discovered by Jeffreys were applied for the first time to a criminal case in England in 1987, in what it is considered the starting point for forensic genetics; the application of individual genetic markers served to both exonerate a man falsely accused of two rapes and killings, and find the true murderer.[14] The victims, two fifteen-year-old girls named Lynda Mann and Dawn Ashworth, were raped and strangled. Initially, a seventeen-year-old boy with learning difficulties from Narborough in the United Kingdom, Richard Buckland, confessed under questioning by the police to Dawn's murder and was charged with the crime. On interrogation, however, he categorically denied being involved in the murder of the second victim, Lynda. Police were convinced both crimes had been committed by the same man due to the similarities between them but couldn't get Richard to confess to the second. Jeffreys got a call from the police asking if he could shed any light on the case with his new genetic technique. He worked through the night with DNA extracted from the suspect and semen found in the bodies of both victims. His results puzzled the police; although they confirmed that both girls were raped and killed by the same man, it wasn't Richard, and the suspect was subsequently released free of charges. Faced with a serial killer in their neighborhood, police chose to make use of the same genetic technology to screen every male born between 1953 and 1970 who had recently lived or worked in Narborough. After eight months, 5,511 men had given blood samples, but none matched the genetic profile found in the semen samples. One of the sample providers, apparently, was a twenty-seven-year-old baker named Colin Pitchfork. But in August 1987, more than a year after the crimes, it was discovered that a friend, on Pitchfork's request, had faked his identity by manipulating his passport. After being arrested and facing the matching of his blood to the crime samples,

Pitchfork confessed to both murders; he was sentenced to life imprisonment. (In April 2016, the parole board recommended, in a rather controversial decision, that he be moved to an open prison.)

And yet there remain concerns and complaints over the use of DNA
profiles in court.[15] For instance, researchers from the Arizona Department
of Public Safety's DNA laboratory reported in 2001 what they thought was
a striking curiosity: by looking at their genetic databases, they found two
unrelated men, one white and the other Black, who matched completely
in nine STR profiles and partially in three more.[16] They could have been
linked to a crime scene where only a partial profile was retrieved from a
problematic DNA sample. The Arizona database contained sixty-five thousand profiles, so the obvious question was whether this match was due to
an incredibly unlikely coincidence or if such problems came up more often
than expected. An additional search found 122 partial matches of different sets of nine STRs in the same database as well as 20 ten-STR matches, 1
eleven-STR match, and 1 twelve-STR match.

The Arizona story was picked up by defense lawyers in 2005. A public attorney named Bicka Barlow who was defending a man in California,
accused due to the match of nine STRs, argued that others could also have
matched, perhaps on the order of 100 or more in the state of California
alone. Bruce Budowle, the FBI's leading DNA scientist, has publicly dismissed such databases as Arizona's for purposes of identifying individuals by kinship analysis on the basis that they were not rigorously curated
and could contain duplications or other problems.[17] The whole story obviously represented a challenge for forensic researchers, who nevertheless
acknowledged that partial coincidences were to be expected according to
probability theory and did not invalidate the general use of forensic DNA
identification.[18]

Individual Historical Genomes

Although there are limitations to working with partial genetic profiles, as
we've seen in the case of the royal French remains, the application of new
and improved genetic techniques to authentic historical samples is revealing new notions of individuality too. To take one case, the genetic analysis
of eight locks of hair attributed to renowned composer and musician Ludwig
van Beethoven (1770–1827) concluded that five of these locks, including

those that showed intact chains of custody, had been derived from a single male—most likely Beethoven himself. This enabled the reconstruction of Beethoven's genetic ancestry, providing a way to examine this figure in unprecedented detail. By combining DNA and archival documents, the researchers also revealed an episode of an extrapair paternity event between Beethoven's lineage and Beethoven's family descendants.[19] As Beethoven never married and had no known offspring, the researchers attempted to match the genetic profile from these hairs to living people with the same family name in Belgium and Austria as well as to living descendants from a common ancestor, Aert van Beethoven, who lived between 1535 and 1609. Yet neither the former nor the latter samples of males matched Beethoven's Y chromosome that derived from an I lineage. An additional search in the FamilyTreeDNA database found six people with closely related—although not identical—Y chromosome profiles to the one found in the compos-er's hair (figure 2.1). The most plausible explanation is that an extrapair paternity event took place between the generations of Aert and Ludwig. There were in fact contemporaneous rumors about the paternity of Lud-wig's father, Johann van Beethoven (ca. 1739–1792), who had a conflictive relationship with Ludwig's grandfather, known as Ludwig van Beethoven the Elder (1712–1773). (Ludwig the Elder's wife, Maria Josepha Poll, was an alcoholic who ended up being placed in the custody of a clinic.) Due to the lack of additional genetic samples to test, the researchers were unable to conclusively determine the solution to this apparent discrepancy between Beethoven's legal and biological genealogy. The members of an association of Beethoven's family descendants from Belgium were not at all pleased with the genetic incongruence. Apparently, the fact that they might still be connected to Beethoven through the maternal line didn't carry enough weight with them. (Certain ancestry companies now offer customers the chance to ascertain if they're genetically related to the musician, although it is unclear to me which genetic lineage they'll choose—presumably his Y chromosome with no known living relatives.)

Beyond the ethical dilemmas involved in the analysis of personal arti-facts from deceased people, as with Beethoven's locks, there can be certain legal implications.[20] In the case of the composer, there is the Heiligenstadt Testament, a letter he wrote in 1802 in which he explicitly asks his doctor to find medical explanations for his myriad ailments. Such research can pave the way for a new field that we can call "genohistory"—that is, the

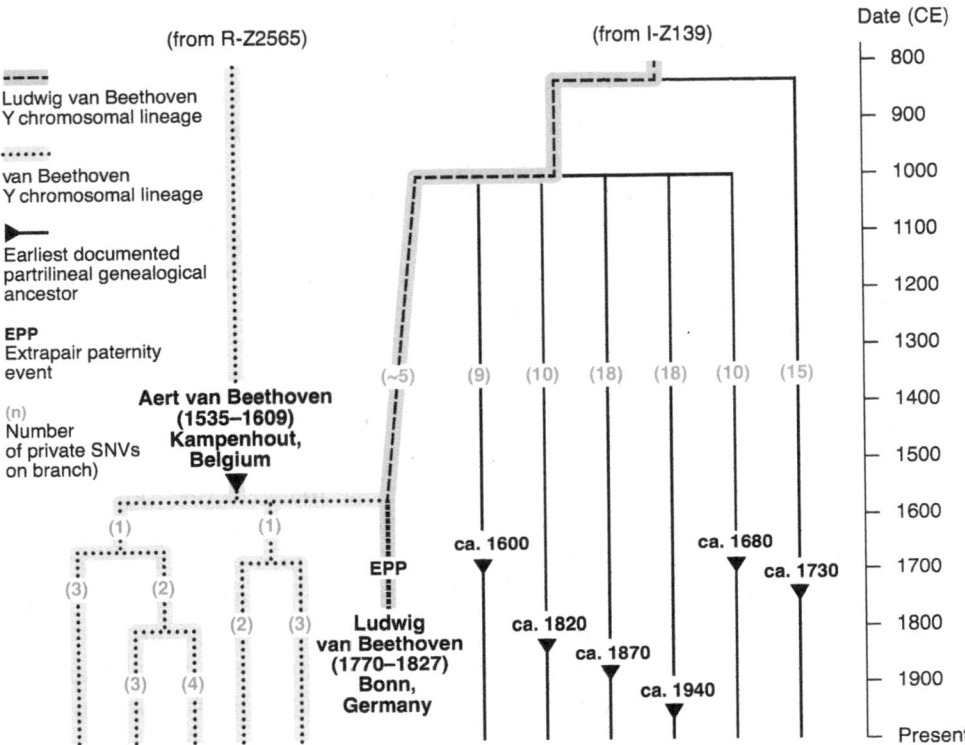

Figure 2.1
A genetic tree of Ludwig van Beethoven's paternal chromosome, depicting the five living Aert van Beethoven descendants, the hair samples, and six unrelated people from the FamilyTreeDNA database that show a closely related Y chromosome to the hair's profile with a common ancestor around a thousand years ago (between brackets, the number of private mutations within each branch as well as the year of the earliest-documented patrilineal ancestor). EPP marks the occurrence of the extrapair paternity event detected. Ludwig belonged to an I Y chromosome lineage (I-Z139), while the descendants of Aert belonged to a different lineage (R-Z2565).

understanding of historical circumstances through the analysis of genomes from their protagonists. Just as our genome determines aspects of our health—and quite probably our behavior as well—the interpretation of personal genetic DNA might even help to grasp decisions made in the past that are sometimes difficult to rationalize.

One consequence of the availability of individual genetic information is the development of a new scientific field, personalized medicine, in which health care resources may be optimally applied to the unique genetic makeup of a particular patient, both for prevention and treatment strategies. In the case of Beethoven's genome, researchers found sequences of hepatitis B virus and also discovered he had genetic risk factors for liver disease, which, coupled with his well-documented regular alcohol consumption, were the likely cause of his cirrhosis. (His last words were "Pity, pity, too late," as he was told of the arrival of a gift of twelve bottles of wine sent by his publisher.) Interestingly, they were unable to find common variants for deafness in his genome. Knowing the gene variants of each individual enables us to predict responses to specific drugs and their side effects and eventually find the most suitable treatment for each case—hopefully to result in more efficient and equitable health care.[21] Thus individual genetic profiles contribute not only to accurate identification but also to better prospects for disease treatment, including earlier disease detection.[22]

Searching for Criminals in the Genetic Databases

Forensic techniques can now identify historical figures or crime suspects from the minutest biological samples. Moreover, genetic data obtained from public genealogical databases to which people upload their ancestral profiles is sufficiently robust to identify individuals not included in those databases. Probably the most notorious example of this approach to identification is the case of the Golden State Killer, linked to at least fifty rapes and twelve killings between 1976 and 1986. The arrest of seventy-two-year-old former police officer Joseph James DeAngelo Jr. on April 24, 2018, many years after his crimes, put the focus on the way he was discovered by the police. The procedure involved searching databases of genetic profiles from private ancestry companies for relatives who could partially match the DNA profile obtained from a sample collected at the crime scene to provide a familial link since the killer himself was not a customer. For decades, the suspect's DNA sample remained unmatched to any genetic profile in forensic genetic databases. Then a retired cold case investigator, Paul Holes, attempted a new approach, searching for potential distant relatives on a genealogy website with genetic profiles uploaded by ancestry company customers. While it is easy enough to find a close relative through DNA similarity, it remains unlikely that one's few close relatives are customers too. On the other hand, chances are that some more distant relatives (say, third cousins, who share only around 1 percent of their kin's DNA) are present in these databases. With information on multiple distant relatives of the Golden State Killer at the level of third cousins, the investigators built up a genealogical tree linking them all going back five generations—in a process they called "reverse genealogy"; after filtering for location, age, and gender, all evidence pointed to a single suspect: DeAngelo. In 2020, DeAngelo received multiple consecutive life sentences without possibility of parole.

This is not the only example of the application of these enormous genetic databases for tracking down murderers in cold cases; since the DeAngelo investigation, sixteen other cases have been solved following genetic genealogy searches, and the practice is on the rise.[23] On November 18, 1987, a few weeks after DeAngelo's arrest, the police managed to identify William Earl Talbott II as a suspect in the double homicide of a young Canadian couple, Jay Cook and Tanya Van Cuylenborg. He was found guilty in June 2019 and received two sentences of life imprisonment without parole.

In that case, the police consulted a genealogy website called GEDmatch to locate two distant cousins who shared the suspect's DNA, found at the crime scene. A reverse genealogical reconstruction showed that the relatives probably shared great-grandparents with the suspect. Again, the genealogical analysis left only one possible male suspect: Talbott.

Nevertheless, ethics experts have expressed concerns over the use of DNA samples originally intended for genealogical purposes for other reasons without the donor's consent.[24] In the case of the Golden State Killer, some customers unknowingly helped the police and genetic genealogists track down a criminal suspect. Only one set of "interim" guidelines, published by the Department of Justice in 2019, regulates the practice, but there is evidence of genetic genealogists skirting privacy rules from DNA database companies in their zeal to solve criminal cases.[25] In some situations, courts have granted law enforcers access to biobanks and databases, but requests have been turned down on occasion. Possible solutions would be to inform future customers of the potential "collateral" uses of their personal genetic data and to improve transparency on the whole for data handling.[26]

This is not the only controversial use of such databases; for instance, thousands of people have discovered through ancestry companies that their parents were not their biological progenitors. Furthermore, the discovery of half-sibs can tell customers that their biological fathers are sperm donors. And although the DNA databases are not intrinsically racialized systems, they are in fact generated by racialized social processes—for instance, Black

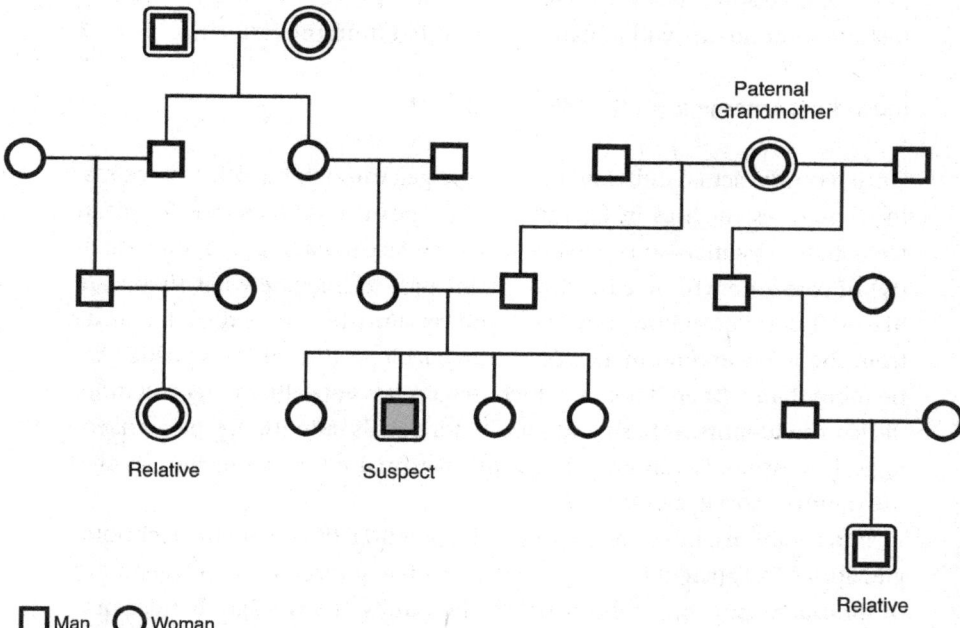

Figure 2.2
Identification of two relatives in the public database (labeled relatives) that pointed to one sole suspect in the Cook–Van Cuylenborg cold case.

males are overrepresented in US DNA banks just because they have higher incarceration rates, which in turn might be a reflection of poverty and disproportionate admission rates.[27] It seems clear that there remain certain legal issues to be settled with this type of information.[28]

Even so, genetic databases continue to grow. The FBI has already amassed 21.7 million genetic profiles, which represents 7 percent of the US population.[29] Sample collection was initially limited to people convicted of crimes, but now the police can take DNA samples from suspects arrested for but not formally convicted of felonies. And in 2019, it was announced that sampling would be expanded to include undocumented immigrants detained by border authorities. Vera Eidelman, a staff attorney at the American Civil Liberties Union, warned that the FBI is getting close to a universal database; this means that it will soon be possible to identify anyone, whether or not they're in the database, with implications for the notions of presumption of innocence and privacy. Unlike the situation throughout most of human history, soon no one will remain disconnected from the rest.

Individuals Emerge from the Mist of the Past

Until recently, actual individuals only emerged through historical accounts. In some cases, such as in Egyptian or European mummies—the Tyrolean Iceman, for instance—it is possible to see the face (a hallmark of individuality) of someone who lived in the distant past to imagine what they were like or perhaps even how they lived, and certainly how they died. But aside from these few exceptional cases, the archaeological record has yielded little more than withered bones, which remain less appealing to us and more difficult to identify as flesh-and-blood individuals (admittedly, bioarchaeological assessments still provide us with personal information such as age, life traumas, social status, etc.).

In recent years, however, thanks to the potential of sequencing technologies applied to ancient bones, it's been possible to reconstruct the genomes of thousands of people who perished thousands of years ago. Besides providing us with a description of their physical traits as deduced from their genomes (the recent analysis of the Tyrolean Iceman, for example, revealed that he had rather dark skin and was likely bald), such studies have managed to identify people's genetic ancestry to a great level of detail.[30] The combination of a phenotypical description along with an ancestry that is sometimes an outlier to the common contemporaneous ancestry offers a vivid perception of individuality previously unknown in the archaeological record.

For instance, in a genetic analysis of hundreds of people buried in the military city of Viminacium, by the Roman Danubian border (today in Serbia), one individual in particular surprised the researchers.[31] He was a fifteen- to eighteen-year-old male, interred with an oil lamp depicting a spread-winged eagle, a bronze coin in his mouth from the first or second century CE, ceramic pots, and a glass vessel—an expensive item. He was buried near a large mausoleum that may have belonged to Emperor Hostilian, an indication of the youth's social status (Hostilian was the son of Trajan Decius and brother of Herennius Etruscus, both of whom died at the disastrous battle of Abritus in summer 251 CE). His ancestry was far from local.

The results clearly pointed to East Africa, as he was most akin to modern Sudanese populations, perhaps in the ancient region of Upper Nubia; his genome indicates he was dark skinned. In any case, this teenager was likely born beyond the southernmost borders of the Roman Empire, ultimately to die by a northern frontier. The eagle in the lamp could have been associated with Zeus, but it was also common in Nabatean culture. This is interesting because the Nabatean trading routes linked the Roman Empire with the Indian Ocean and perhaps the youth had resided somewhere along the Red Sea route (an isotopic analysis of his bones indicated marine protein among his nutritional sources). Though we may never know his name or personal details, scientific evidence of his incredible journey—a socially important aspect of his personal identity—is now public knowledge. Somehow, this person has been returned to human history.

This is not the only example of long-distance migration within the Roman Empire. Near the northernmost border of the empire, in ancient Eboracum (in modern York), a Roman necropolis was discovered and excavated in 2004. The skeletons dated from between the early second century and the later fourth century CE, and most seem to correspond to gladiators—tall, well-fed young males who display abundant signs of trauma as well as some bite marks that could be attributed to large animal, such as bears or even lions. Some had been decapitated, with the head lain between their legs or on their chest, possibly a procedure to ensure they would not return from the grave to pester the living. The genetic analysis of a bunch of these skeletons revealed that few were from Britain itself and some came from continental Europe. One individual, however, hailed from much farther away; his ancestral composition points to modern-day Jordan or Syria as his likely place of origin. We'll never know the reason for his surprising journey that started in the Near East and ended up in cold, distant Britain, nor for his headless state at burial.[32]

It can be argued that the Roman Empire represented the first case of globalization, as it affected regions on three different continents. But there are other examples, some quite intriguing, of long-term life journeys in pre-Roman times. In ancient genomic research on prehistoric samples from the Iberian Peninsula, we found a male, dated to 2500–2000 BCE, whose DNA showed affinities with that of modern North Africans. His ancestry was different from other Iberian Copper Age individuals analyzed. The most surprising part of the story is that he was buried at a site called Camino de las

Yeseras near Madrid (Spain); that is, he was far from the coast—and North Africa. Interestingly, there were many ivory findings at the site, which may suggest a role for an African; in this case, he might have been a trader on long-distance Copper Age commercial routes.[33] As in the case of Viminacium or York, we can only guess how his foreign origin—and most likely, contrasting physical appearance—affected his notion of personal identity in the social context where he died.

These genetic studies are providing a new picture of individuality across prehistory beyond the historical record. Even if these individuals from the past remain—and will ever remain—anonymous, the extraordinary level of detail within their genetic information can present an image of them as individuals, setting them apart from their contemporaneous peers. Moreover, it is likely that some of these individuals won't remain unconnected to the rest of the human family for long. If they had any descendants, there is the possibility that some will be genetically linked to others—even in our own era, as more and more ancient genomes closer in time to us are generated. Ancient individuals are emerging from the mists of oblivion to tell us they existed. And soon they will be connecting with each other and us. In many ways, it is also a democratic revolution.

Genomics reminds us of the unique dimension of individuality. Even in a world that's now populated by a staggering eight billion inhabitants, each person has a unique combination of genetic variants. If such diversity is a sign of identity, then we can conclude that we are unique indeed. The question remains how to translate this inconceivably huge quantity of genetic information into any sort of interpretable notion of uniqueness and identity. Most of the genetic variation so far described bears no clear phenotypical impact that might help us figure out the ultimate meaning of our personal genome. Unraveling and interpreting these complex genotype-phenotype relationships as they bear on personal identity issues are formidable scientific tasks fraught with numerous uncertainties. Still, as science progresses, so too will the physical and even behavioral aspects of our individuality emerge and link to our genome. It will be the culmination of decades of research in the deep understanding of our genetic individuality.

3 Alikeness: Twins and Doppelgängers

We're not searching for anything except people. We don't need other worlds. We need mirrors.

—Stanisław Lem, *Solaris*

In 1992, when I was in my late twenties, I attended a meeting on human evolution in Jerusalem. One day, walking in the street, I was approached by a young lady with long and dark curly hair who apparently knew me and was clearly happy to meet me. She got very close and started talking to me in Hebrew with a big smile; all I could do was tell her to speak to me in English. At first she was puzzled and then infuriated. As I kept saying I could not understand Hebrew, she finally left, twisting her finger at her temple in the universal gesture for insanity. I assumed I had a double somewhere in Israel, someone who obviously looked very much like me. I sometimes think about this ghostly figure I never met. Does he still have curly brown hair, as I once did, or is he getting bald, as I am? What about the life he leads? Did he convince her I was another person? Maybe they are asking themselves the same questions from time to time.

In searching for shared identity, one might think the most indicative feature of potential alikeness, beyond such general traits as pigmentation, would be facial similarity. Faces are indeed the most distinctive feature of our unique individual identity and represent the external manifestation of ourselves. Facial recognition involves sophisticated brain machinery; it seems we've been under evolutionary pressure to look unique and thus easily recognizable in a crowd.[1] The central role of faces in identity explains why identical twins get our attention or the psychological problems that certain disfigured people undergo in gaining a new identity after being subjected to facial allograft transplantation.[2]

Identical twins are not such a rare phenomenon. In fact, they account for about 1 in every 250 live births.[3] I have my own identical twins' story. Some years ago, I went to Copenhagen to visit a well-known paleogenetic researcher, Eske Willerslev. We went to a pub to chat over a few beers. After a while, I needed to go to the rest room, but when I got back to the table, I discovered what seemed to be a perfect copy of Eske sitting beside him. I thought for a moment I was utterly drunk before I realized he had an identical twin, Rane Willerslev, a renowned Danish anthropologist, who has just popped up. Even their voices sounded identical (interestingly, this is partly a subjective perception because automatic speaker recognition systems generated only 2.4 percent of twin pairs voice misidentification).[4]

I have two brothers, but we are quite different. I always wondered what it would be like to have a person around who looks similar to you in a mirror. I guess this feeling is expressed in a memorable sentence by Argentinian writer Jorge Luis Borges (1899–1986): "Mirrors and copulation are abominable, since they both multiply the number of men."

Twins in the Past

Twins can now be screened at archaeological sites thanks to genetics. The earliest identical twins, found in the Upper Paleolithic double burial of Krems-Watchberg (Austria), were infants; one died at birth, and the other survived for about fifty days. Interred in a Gravettian archaeological context and dated to thirty-one thousand years ago, they were buried side by side, along with some ornaments such as mammoth ivory beads, a perforated fox incisor, and three perforated mollusks.[5]

These aren't the only potential twins found in the archaeological record, though discussion of other cases, lacking similar genetic confirmation, remain inconclusive. But unless both died at birth, finding twins in the prehistoric record is unlikely; in traditional societies, such as the Aché or !Kung, it is said that the births of twins often implied the killing of one of them because of the impossibility of feeding two kids simultaneously.[6] Prospects for twin survival were grim if their own existence was intermingled with religious rituals. In a recent analysis of sixty-four skeletal remains from a sacrificial cave near the Sacred Cenote in the ancient Mayan city of

Chichén Itzá (modern Mexico), the genetic researchers discovered that not only were all the victims boys but also two pairs of identical twins were among them. A potential interpretation suggested they were selected for sacrifice because of the importance of hero twins Hunahpu and Xbalanque in Mayan cosmogony.[7]

Some twins in the past survived to adulthood. Especially intriguing are two males buried in close proximity to Kuline cemetery at the abandoned Roman fortress of the Timacum Minus site in present-day Serbia. They are dated to the tenth century, and their genetic ancestry matches that of current southwestern Europe. Although it's impossible to pinpoint their precise place of origin (it could be southern France or Iberia), what is surprising is that they died thousands of kilometers to the east in the Balkans. The fact that they were buried side by side means others were aware that they were close kin, although it is impossible to know if they died at the same time.[8]

Identical twins, despite the common name, aren't entirely identical. Siblings usually differ in subtle physical traits as well as in susceptibility to some diseases, notably such neurological conditions as autism, bipolar illness, and schizophrenia. While these differences traditionally have been attributed to the environment, an obvious additional hypothesis to test, with the advent of genomic techniques, is the effect of genomic and epigenomic differences that could have occurred after zygotic formation.

The results indicate that identical twins are not always genetically identical; some mutations might happen in early developmental stages (when different cell types are being formed) in one twin or another. In a recent study, researchers wanted to address the number and type as well timing of the occurrence of mutations differing between identical twins.[9] They sequenced 381 pairs of twins—plus 2 triplets—in several somatic tissues (mainly in blood samples and buccal swaps) and found a total of 23,653 postzygotic mutations in the sample, with an average of 5.2 mutations differing between twin couples. The distribution of these mutations across the sample was quite variable; some pairs (38 of them) were effectively identical while others (39, or approximately 15 percent of the sample) differed by more than a hundred mutations. Also, the majority of differences were found while comparing blood samples, suggesting that clonal hematopoiesis—the process by which groups of blood cells are formed within the body—is prone to generating mutations. While somatic mutations are generated throughout an individual's life and thus tend to accumulate with age, the researchers discovered that most observed differences between pairs of twins took place in the early developmental stages—indicating they were not a function of the age of the individuals.

The differences can be read not only at the DNA level but also in the chemical modifications within the DNA itself that regulate the expression of genes by silencing or enhancing gene activity—that is, the epigenome. Analysis of the epigenome in twin siblings has shown that at least in one-third of them, there are in fact differences in the patterns of both methylated DNA regions and histone modifications differentially affecting the expression of some genes in some tissues. These chemical signals can be influenced by both external and internal factors, and tend to diverge with age, which could explain why older twins are less alike than younger ones.[10] Some habits, including diet, smoking, or physical activity, have been proposed to induce epigenomic modifications that can be inherited over several generations. Such differences would be more evident in those twins who have spent time apart or have different lifestyles. The result would be to bring an almost identical genotype into slightly different phenotypes, including physical appearance as well as late onset diseases.

The genomic results suggest that while physical differences between pairs of twins can be attributed mostly to differences in the environment, at least in a significant percentage of the cases there are a substantial number of genetic and epigenetic differences between twins that can play a role too. This could happen, for instance, in cases where one twin has a health condition, as it is known that new mutations as well as epigenomic modifications influence the emergence of some disorders.

Two People in a Single Body

While twins represent a challenge to our perceptions of individuality, what then can we make of the existence of conjoined twins (or Siamese twins)—that is, twins born physically connected to each other in varying degrees? Their incidence is difficult to estimate, although it could range between as high as 1 in 50,000 births to 1 in 190,000; at least in Western countries, they are increasingly rare due to the availability of pregnancy monitoring. They are originally identical twins whose embryos split later than the eight to twelve days after conception that ordinarily takes place in the separation of identical twins and hence separation of the siblings halts before the process is complete.

In the past, conjoined twins possessed a peculiar fascination, particularly during the Middle Ages; they represented a moral dilemma as well, perceived as they were as contrary to nature—accidents that demanded an explanation. In 1429, the birth of conjoined twins in Aubervilliers (France) provoked such curiosity that chronicles say about ten thousand people went from Paris to that provincial town to see the phenomenon. And the Scottish Brothers, born in Glasgow around 1460 with two heads on a single body, become popular in the court of the Scottish king James III until their death around 1488.[11] It seems they performed duets—one singing tenor, and the other bass—to entertain courtiers. According to contemporary accounts, they would argue bitterly and even physically attack each other.

At the beginning of the nineteenth century, embryologists began to systematize different types of teratologies. The founder of rational teratology, the scientific study of malformations—or "monsters," as they were known at the time—was Étienne Geoffroy Saint-Hilaire (1772–1844). His work was carried on by his son Isidore (1805–1861). Saint-Hilaire classified conjoined twins according to the zone of union of both twins. For instance, he called them "thoracofagus" if they were joined at the chest. He further classified them as symmetrical and asymmetrical, the latter term referring to twins that differ in size, with one sibling—also known as a parasite— much smaller than the other. A famous example of such asymmetry was Lazarus Coloredo and his parasitic brother, John Baptista, who consisted of one head, two arms, and one leg that hung from Lazarus's navel. Coloredo arrived in London in 1637 to present himself, with great success, at an exhibition. By 1642, he was still alive and touring Scotland.[12]

Undoubtedly, the most famous conjoined twins were Chang and Eng Bunker (1811–1874), originally born in Meklong, in what was then the kingdom of Siam (now Thailand). The term "Siamese twins" may be attributed to their popularity. Their only point of union was a section of body that shared a fused liver; Chang was positioned on the left and Eng on the right. After emigrating to United States, they performed at exhibitions for a fee, eventually settling in Mount Airy, North Carolina, where they married two sisters and established two large families (Cheng had ten sons and Eng twelve). They owned a plantation and a number of slaves, who

Figure 3.1
A photograph of the conjoined twins Chang and Eng Bunker, born in Siam (now
Thailand). They were so popular in United States that the phrase "Siamese twins,"
used as a generalization for conjoined twins, derived from them. *Source*: Wikimedia
Commons.

were emancipated after the Civil War (the brothers apparently supported the Confederacy). They died at sixty-two—until recently, the longest-living conjoined twins.[13] An autopsy confirmed that any attempt at a surgical separation, an option that the brothers had consulted numerous doctors about throughout their lives, would have been fatal.

Having seen the physical similarities between genetic twins, we can easily understand how they might be perceived as sharing a single identity. But how about physical similarities between those who are not twins or not even related? Do they too share an identity?

Stories of Our Doppelgängers

In recent years, finding look-alikes, or doppelgängers (a German noun formed by the words *Doppel*, double, and *Gänger*, walker), has become a popular pastime on social networks. The idea of such existing doubles—that is, someone who looks like someone else but is unrelated—has fascinated humans since antiquity, although it has usually been cast in a ghostly light or attributed to a bad omen. Writers and filmmakers have explored the possibilities of discovering copies of ourselves, without previous knowledge of their existence, that look identical and yet are different people. In general, they've emphasized that looking alike does not mean being the same person.

Doubles can have moral overtones. In one of the first examples of doubles in a modern narrative, that of William Wilson, created by Edgar Allan Poe (1809–1849), the narrator—not the double—is morally defective. In this tale, first published in 1839, the double represents the good that remains in the soul of the narrator, who, in his descent into depravity, ends up killing his good version. The story starts at school, where there is a boy who has the same name as his own, William Wilson, despite being unrelated to him (later on, the narrator notes that they were born on the same day). The second Wilson attempts to stop the original from doing as he pleases, but always when they're alone and no one else can see or hear them. (One wonders if this is all in the imagination of the original William.) As they grow older, they remain alike, "but I saw that we were of the same height, and I perceived that we were even singularly alike in general contour of person and outline of feature." They have only one perceptible difference: the double can only speak in a soft, low whisper. As in other doppelgänger literature, other people seem not to be aware of the resemblances. "But, in truth, I had no reason to believe that . . . this similarity had ever been made a subject of comment, or even observed at all by our schoolfellows." The copy keeps telling the original what he should and should not do, unsuccessfully trying to restrain him on his path to wrongdoing. The narrator leaves the school, only to find, three years later, his William Wilson once again at Eton, where the original indulges in heavy drinking and card playing. Leaving Eton, he takes refuge at Oxford, persisting in his dissolute ways. There he meets a rich young man and immediately plans to trick him at cards. Having duped the poor gentleman, who loses everything in a few hours, someone dressed like the original Wilson and speaking in his characteristic whisper bursts in, informing those in attendance of the deception. The narrator, who is expelled from Oxford due to the scandal, is astonished to notice that the intruder Wilson had left behind an exceedingly rare cloak like his own, thus revealing the exactitude of the copy, even in what he wore. Finally, after many similar episodes over the years, the original finds the second Wilson during the Carnival of Rome and enraged, impulsively drags him into a duel, plunging his sword through the man's bosom several times. The dying Wilson then utters his final words, now in his own voice: "You have conquered, and I yield. Yet, henceforward art thou also dead—dead to the World, to Heaven, and to hope! In me didst thou exist—and, in my death, see by this image, which is thine own, how utterly thou hast murdered thyself."

A similar idea can be found in Fyodor Dostoyevsky's (1821–1881) *The Double*, a short novel published in 1846; Golyadkin Senior, a rather antisocial and insignificant office worker, meets someone in a snowstorm who looks like him and even has the same name—referred to as Golyadkin Junior in the novel. His copy is a friendly and charming person with all the social skills that the original lacks, and he grows progressively more popular at the original's workplace, where colleagues are apparently unable to notice their similarity. Golyadkin Senior increasingly thinks the other is leading a conspiracy against him, humiliating him in even the smallest daily details. In a restaurant, for instance, the original has a pie, but when he attempts to pay for it is told by the waiter to pay for eleven pies, as that is how many he's supposedly eaten. Astonished, he tries to argue with the waiter, only to glimpse his double somewhere else in the crowded restaurant, grinning and nodding toward him. In the end, an embittered and increasingly confused Golyadkin Senior goes to the home of his superior—called "his Excellency" by everyone—to denounce what he considers persecution against him by

Golyadkin Junior. As a servant attempts to block his unexpected visit by the door, he spots the double with his Excellency, apparently about to leave his host on friendly terms. Afterward, Golyadkin Senior has a nervous breakdown and is taken to a mental asylum, leaving the reader to wonder if his charming, socially successful counterpart is nothing more than a product of his imagination.

Science fiction writers have used the notion of copies too. In "The Seventh Voyage," a short story included in the *Star Diaries* by Stanisław Lem (1921–2006), the doubles are our own selves from the past and future (Are we really the same person?). The story tells the tale of a rocket that's been hit by a small meteor, making it unsteerable, and it can only be repaired by two astronauts in space suits. Realizing this, the sole crew member and narrator, Ijon Tichy, goes to sleep feeling hopeless. The rocket then undergoes several temporal loops in which successive copies of the astronaut appear. One of Tichy's replicas attempts to awaken him and urge him to get to work, but Tichy dismisses him as an illusion. Over the next few days, more and more copies are generated but nothing is accomplished. Some of them conclude that there's no point in doing anything since whatever efforts previous copies made in the past were insufficient to repair the rocket and send it in the right direction. Copies accumulate inside the cramped spaceship as it goes through subsequent time warps and conflicts arise. "When I regained consciousness, the cabin was packed with people. There was barely elbow room. As it turned out, they were all of them me, from different days, weeks, months, and one—so he said—was even from the following year. There were plenty with bruises and black eyes, and five among those present had on spacesuits. But instead of immediately going out through the hatch and repairing the damage, they began to quarrel, argue, bicker and debate." The crowd of Tichy's copies decide to form a committee to make decisions, but in traversing a negative vortex half of them disappear. Now lacking a quorum, they need to decide on new voting rules, and the debates and disagreements go on as new vortexes continue to decimate the crowd. In the end, the original Tichy awakens alone to find that two child versions of himself, using the sole remaining space suit—one putting his arms in the sleeves, and the other in the pant legs—have managed to go outside and fix the problem. The message is that the boys are the only ones able to take action, demonstrating a resourcefulness that the older doppelgängers lacked.

Doppelgängers, it appears, can be quite effective as fictional characters. But daily experience tells us there are people who do indeed look quite similar, and this can have consequences—in justice, for example, where facial identification invariably serves as key evidence. There's a popular story about two Black American inmates, both named William West, who looked similar enough to be twins. The second West was brought to the Leavenworth Penitentiary in Kansas in 1903. During the photo session, a prison official was baffled to see someone he recognized, thinking he'd seen the man arrive at Leavenworth two years earlier to serve a life sentence, despite the new arrival denying he'd ever been there. After searching the archives, they discovered another inmate who looked exactly like the new arrival, also by the same name. Though unrelated, the two bore a striking resemblance, even measuring the same height. The case is said to have forced the justice system to seek more reliable methods of identification, and in fact fingerprinting came into use just a few years later. This was no unique case; misidentification by witnesses still goes on today. In 2017, a man who spent seventeen years in prison, Richard Anthony Jones, was released when it was discovered that his doppelgänger, Ricky Amos—another case of surprisingly similar names— might be the one responsible for the robbery he was accused of. Photos show that the two did indeed bear a striking resemblance—in this case accentuated by matching goatees and long hair pulled back in similar fashion.[14]

Individuals may be subject to such confusion, and not just because of their appearance but also because of their names (another sign of identity, although in this case merely cultural). Recently, writer Naomi Klein wrote a book about being mistaken for another writer, Naomi Wolf—a phenomenon that illustrates the "instability of identity in the virtual world."[15] Klein concludes that doppelgängers in the modern world can have a political dimension, with far-right doubles such as Wolf being a distorted image of the left social ideology.[16] What can genetics tell us about modern, physically similar doppelgängers? Is there a scientific explanation for them?

Genetics of Doppelgängers

The expansion of social networks in the twenty-first century has enabled the spread of literally trillions of pictures of humans across the globe, and with it the discovery of an increasing number of "virtual twins" as they're called despite an absence of family ties. Examples of such pairs can be found online at a website documenting a project initiated by Canadian photographer François Brunelle in 1999.[17] (A variation involves discovering old photographs, sometimes from as far back as the mid-nineteenth century, of people who look quite similar to present-day individuals. Some of the striking examples that come to mind include an 1870 doppelgänger of Nicholas Cage and one from 1860 that was a dead ringer for John Travolta.)

Such similarities aren't just a subjective perception; there are in fact ways to quantify the phenomenon. Aside from the comparison of biometric data between pairs, perhaps the most powerful way to test the resemblance between look-alikes is to use facial recognition software. Now widely used in security surveillance, it provides a facial similarity score that ranges from zero to one, with the higher figure indicating identical faces. In about half the look-alike pairs, the computers considered them the same individual, while the rest got scores close to one. The confounding effect on visual recognition that many of these couples produced was on the same level as that of identical twins.

In a pioneer study on the genomics of doppelgängers, a group of geneticists—myself among them—decided to investigate if such look-alike couples could have similarities at the genetic level or unrecorded kinship links, or both.[18] They contacted people portrayed in Brunelle's project and obtained genetic consent from thirty-two of them—that is, sixteen look-alike pairs—and subsequently sequenced their genome, epigenome, and microbiome. That done, the researchers used algorithms commonly employed in phylogenetic reconstruction to generate a tree where genetically similar individuals tended to branch together. They found that in more than half the cases, both members of the pair were clustered together, indicating a genetic similarity. At the same time—considering that first-degree relatives (parents-offspring or brothers-sisters) share 50 percent of their genomes, second-degree relatives 25 percent, and so on—they were able to discard any close family relationship between the pairs. That is, though the doubles had similarities at the genomic level, there weren't enough of these for them to be classified as relatives. By contrast, the epigenome and microbiome registered no noticeable similarity between the pairs, suggesting that these factors were not involved in facial resemblance.

It was found that the nine pairs with the closest resemblance had the same genetic variants in 19,277 positions in the genome, affecting 3,730 genes. This similarity was larger than what would be expected in randomly selected couples from a genetic dataset of the same human population. Of course, we can guess that such sharing might have implications in the phenotype, and indeed researchers found an enrichment for identical variants in genes associated with anatomical structure development and morphogenesis. (It's worth keeping in mind that we still don't know the precise function of many genes, but we can at least trace their effect on specific body structures or tissues with different functional techniques.) Significantly, some of the shared genes were described to be involved in facial parameters and thus it's reasonable to say that the look-alike pairs had similar facial features because the individuals happened to share a number of genetic variants in a set of crucial genes. And yet it must be stressed that the couples differed in many other genes, which means they're separate enough not to be kin related, at least to a close degree. (A distant genealogical relationship might be difficult to discover beyond, say, the seventh degree because they share a small fraction of DNA.)

Even more interesting, the results indicated that doppelgängers shared, beyond genes involved in determining physical traits, some genetic variants linked to behavioral aspects that could also indirectly influence those similarities. That is, similar habits and lifestyle options (for instance, smoking, diet, or educational achievement) could enhance their ultimate resemblance—just as with actual twins.

It is worth mentioning that this study has been criticized in some methodological aspects, especially for the lack of an appropriate control sample—that is, pairs from the same populations with no obvious physical resemblance; moreover, the most important question about doubles remains unanswered. Why do such pairs exist? An analysis of ancestry demonstrated that the pairs tend to have a similar geographic background. With a few notable exceptions—where, for instance, one member came from Germany and their counterpart from southern Italy—most of the pairs came from the same country. They therefore shared a common populational background at the outset, to which additional genetic variants happened to provide physical resemblances. As the genes involved in facial shape are finite, it is likely that the potential combinations having an impact on physical traits are limited and partially constrained by developmental factors too. Hence the combination of the enormous population of all humankind plus the widespread use of social media—along with our tendency to detect patterns—simply favors such discoveries.

Genetics of Human Faces

The root of the problem is again related to physical appearance and the large variation that exists in human faces, which are more individually distinctive than for most animals. Some genomic studies found that human facial traits (such as nose length and width) seem to be less correlative than other bodily features (for instance, hand length and width), which in turn would suggest that facial diversity has evolved to signal individual identity.[19] While facial recognition is an important social trait in primates, it is likely that the remarkable level of human variation was selected for over millions of years of hominin evolution, in parallel with an increase in cognitive abilities and complexity of social interaction.

In the past few decades, information technologists have developed methods to ensure facial recognition, facilitating identification of human faces by means of a set of biometric measurements and mathematical algorithms. In authoritative countries where privacy is not a priority, it's been developed as a form of social control and surveillance, though it can also have huge implications for marketing.

Facial recognition software captures the shape of one's face by measuring a set of key aspects, such as the distance between the eyes, width of the nose, and distance between the eyeline and chin—measurements that can be made from videos of street scenes. This metric signature of people's faces is then compared to a database—the larger the better for potential identification within a specific population. For 2021, it is estimated that police databases (drawn, for example, from government surveillance at all US airports) contain the facial images of at least 117 million people. Although facial recognition systems now operate at the level of human performance, there have been instances where neither human nor machine succeeded, as in the case of identical twins. Thus identical twins are routinely employed to improve the discrimination of facial recognition software.[20]

We might well ask ourselves how probable it is to find people who look alike and also pose problems for facial recognition systems. Considering that humankind now exceeds a staggering eight billion people across the planet, it could be just inevitable. The first step to estimate the possibility of having at least one doppelgänger like the one I might have in Israel would be to understand the genetic architecture underlying the shape of human faces and their variation. Although this is obviously a highly complex trait, the number of genes involved, though huge, is finite. For instance, in a survey of 2,329 persons of European ancestry, the authors found 38 genetic targets, mostly operating during embryological development and affecting human facial shape in eyes, nose, chin, and mouth.[21] Certain studies register little variation in many of the genes involved in face morphology, which is understandable considering that there's scant room for genetic novelty with a feature that plays such a crucial role for survival (thus the diagnostic function of facial patterns for certain genetic diseases such as Down, Angelman, or Williams-Beuren syndromes). It seems that most variation in human facial morphology hinges on DNA variation or the regulatory sequences of the approximately 4,000 genes involved in craniofacial development—to a greater extent than mutations in the genes themselves.[22] Interestingly, a recent study added a new layer of complexity by suggesting that increased nasal height—our nose is a prominent facial trait—can be attributed to a particular genetic variant inherited from Neanderthals.[23] If, according to Saint Jerome (347–419), "the face is the mirror of the mind," it is perhaps surprising to discover that our face can be the reflection of humans who no longer exist.

One day, Marc, my then eight-year-old son, was running across the house with a soccer ball; when I asked him what he was doing, he answered, "I am playing against myself and right now 'the other' is winning." I think this sentence captures the mysterious philosophical question of having a double. In film and literature, doubles are identical—even if their resemblance is not equally perceived by outsiders—but often morally opposed. That is, they are subtly imperfect copies of the original in moral terms; discovering one's copy usually turns out to be a bad omen. But such potential differences underline the subjectivity of conflating appearance with moral character. People might look identical yet remain essentially different. If anything, looking at our doppelgängers can be used to know ourselves better.[24] Therefore resemblance cannot be taken literally as a signal of identity; neither the face nor the genome is a measure of any innate similarity. It's safe to conclude that however similar we look, we are different. Each twin, each doppelgänger, even if identical, is still unique. In fact, even we, in different moments of our lives, are different people; this is the message of the tale "The Other" by Borges, where one morning the old narrator encounters a younger and still idealist version of himself. Having or not having a double, unknown or known to us, we can conclude that human beings have no genuine copies.

4 Battle of the Sexes

It is fatal for anyone who writes to think of their sex. It is fatal to be a man or woman pure and simple; one must be woman-manly or man-womanly.
—Virginia Woolf, *A Room of One's Own*

The Vikings—legendary Scandinavian raiders, seafaring warriors and explorers, and farmers too—plundered northern and central Europe for almost three centuries, from the eighth to the eleventh century CE. From a modern perspective, the Vikings are in many ways social icons who populate epic images, including TV series and films. The graves of Viking warriors can be found near a number of settlements throughout Scandinavia. In Birka alone— once a key trade center located in east-central Sweden that connected commercial routes as far as the Ural Mountains, the Byzantine Empire, and the caliphate—there are over three thousand graves, more than a thousand of which have been excavated. One particular grave, labeled Bj 581 and placed on a prominent terrace overlooking the town and sea, was especially furnished with funerary objects that include a sword, axe, spear, arrows, battle knife, two shields, and two horses. These were accompanied by a full set of game items that included three antler dice and twenty-eight playing pieces. When excavated back in the 1930s, it was unanimously considered to belong to a prominent Viking warrior. But unlike the funerary objects, the skeleton was poorly preserved and overlooked for several decades.

It was not until 2016 that a full examination of the skeletal evidence suggested the Birka warrior was in fact a woman.[1] This revelation triggered a whole debate about sex and gender roles in the Viking world, with some scholars being reluctant to acknowledge the participation of women as warriors. In 2017, two samples, one from a canine tooth and another from a humerus, were subjected to DNA analysis.[2] The results confirmed it was a woman due to a lack of any evidence of a Y chromosome among the millions of genomic sequences generated. What is interesting about this story is that the male sex attribution was taken for granted; only once the skeleton was known to be a woman did some academics start questioning if the accumulation of weaponry was indeed a sign of warrior identity or something else. Some even suggested there might have been a male skeleton within the tomb that did not survive. Such a question is never raised when the subject is male.

With basic chromosomal information, we are now able to determine the genetic sex of thousands of individuals from the past, even from minimal dental or skeletal remains. Knowing an individual's sex is important for interpreting the archaeological record, prone to prejudice and misconceptions as it is. To take one example, the analysis of a skeleton from a nine-thousand-year-old burial at the Wilamaya Patjxa site in the Peruvian Andes that displayed numerous projectile points and animal-processing tools demonstrated once again that the interred person was female.[3] Based on such evidence, some academics have tried to challenge the traditional sexual division of labor in prehistory, with the perception of man as hunter and woman as gatherer.[4] These investigators, however, tended to select anthropological literature in which female hunting was highlighted while dismissing or ignoring the evidence that female involvement in hunting—especially of large prey—was minimal or absent.[5] This attempt at Western-centric revisionism on ethnographically well-established gender roles is but one instance of the current battle of the sexes in the North American academy.

Sex Determination

The mainstream life sciences explain biological sex as a binary trait defined by the gametes you produce. There are only two types of gametes: one small and usually mobile (sperm, produced by males), and another large and immobile (eggs, produced by females). With few exceptions, all sexually reproducing plants and animals follow this binary rule.[6] The production of gametes correlates with chromosomal architecture. In the case of humans, you are either female with a pair of X sexual chromosomes, or male, with a single X and the much smaller Y chromosome. This is what we call the sex

Figure 4.1
The Viking woman warrior at Birka (Sweden). Drawing of the tomb excavated by
Hjalmar Stolpe, by Evald Hansen, 1889. *Source*: Wikimedia Commons.

determination system. There are downstream consequences of biological sex in morphology, physiology, and even behavior.

Sex differentiation—the developmental processes leading to the expression of a biological sex—is more complicated than the presence of a specific combination of sexual chromosomes or gametes. Scientific evidence indicates that the default sex is female and the presence of a Y chromosome is what turns an otherwise female embryo into a male. Five weeks into development, an embryo can still develop into a male or female. A developing testis starts secreting a sex-associated hormone, testosterone, which triggers the development of male ducts. If the gonad develops into an ovary, it secretes another hormone, estrogen. The sex-associated hormones support the development of external genitalia and will act again during puberty to trigger the appearance of secondary sexual characteristics. In males, these include an increase in bone structure and muscle mass along with changes in the vocal cords and the appearance of facial hair; in women, this includes an increase of body fat in the hips, buttocks, and breasts, among other traits.

At the beginning of the twenty-first century, however, the idea of a passive default femaleness in human sex determination was challenged by the discovery of genes such as WNT4 that actively promote ovarian development, while at the same time suppressing the testicular development. WNT4 is expressed in the undifferentiated gonads even before sex determination; it is involved in multiple developmental processes, including kidney development, and required for proper male and female sexual development.[7] These and other discoveries have established the idea that sex determination is a balance between opposing levels of expression of pro-testis and pro-ovary factors, which include both genes and epigenetic regulations.[8] This balance is still not well understood, but according to some scientists, it can shift long after development is over.

Some academics have argued that gametes alone cannot adequately define biological sex and that this trait is not binary but instead encompasses a complex spectrum.[9] Yet the proponents of biological sex as nonbinary conflate the variety of manifestations of secondary sexual traits with the gamete production that has traditionally defined biological sex. If we accept a fluid definition of biological sex, it becomes difficult to interpret the diversity of the natural world, including what evolution tells us about it. The evolution of sex is arguably one of the most important topics in evolutionary biology and has been investigated by such scientists as Charles Darwin, Ronald Fisher, J. B. S. Haldane, Maynard Smith, Theodosius Dobzhansky, George C. Williams, William Hamilton, and Robert Trivers.

Those who support the notion that biological sex is fluid seem to adhere to a tradition of postmodernism that can render the concept useless as a scientific tool. Although its proponents maintain that the notion of a binary sex is a political and discriminatory supposition, it is likely that denial of this basic biological rule can have negative intellectual consequences. In an extreme scenario, it might trigger the erosion of scientific trust in evolution and, at a moment of climatic emergency, also lend credence to popular views that humans exist apart from the natural world.

Nevertheless, we can ask ourselves if biological sex is the only legitimate description of sex identity in humans, and whether this basic concept, as expressed on our passports, can have harmful social implications for some collectives. I think it is possible to acknowledge the evolutionary mechanisms of sex in nature while at the same time recognizing the complexities associated with human sex identity on multiple levels. And of course I accept some people might consider biological sex irrelevant for their personal identity.

Complexities around the Sexual Chromosomes

To define biological sex at the basic gametic level is not to deny that there are variations in the common sex chromosomal rules in our species. Sex chromosome conditions refer to variations in the number of chromosomes related to sex differentiation: there are females with just one X chromosome (a condition known as Turner syndrome), females with three X chromosomes, males with two or three Y chromosomes, and males with two X chromosomes (a syndrome known as Klinefelter generally resulting in poorly developed testicles and infertility). This is not just a phenomenon of present times; with enough genetic information, it is possible to uncover people from the archaeological record who suffered from the loss or gain of one or more sex chromosomes. In a study of ancient genomes from Iceland, for instance, it was possible to determine that a pre-Christian individual who died before adulthood had Kinefelter syndrome.[10] Another study found an individual from Somerset (England) who lived twenty-five hundred years ago and had a single X copy instead of two in some of her cells.[11] The existence of such chromosomal variations does not invalidate the binary nature of biological sex any more than the existence of people with Down syndrome, who have an additional chromosome 21, invalidates the rule that our species has twenty-three pairs of chromosomes.

The complex relation between sex determination and sexual identity has become apparent in recent years with DNA sequencing technology and cell biology increasingly providing lines of evidence. Such research has also been instrumental in eradicating the social ostracism of people affected by certain conditions.

Certain additional sex issues are raised by the sexual chromosomes themselves, especially the Y chromosome. For several reasons it is a rather odd chromosome, loaded with many repeated—and on occasion even palindromic—DNA segments. It wasn't until August 23, 2023, that its complete, gapless sequence was published, correcting multiple errors in the previous version, including the addition of 30 million nucleotides to the reference.[12] With the current length of 62,460,029 nucleotides, the Y chromosome is among the shortest of the genome—only chromosomes 20, 21, and 22 are shorter. What makes the difference is the number of genes. At the Y chromosome, there are only 27 protein-coding genes—some of them with multiple copies. In contrast, the shortest autosome, 22, contains between 500 and 600 genes.

The evolutionary history of the Y chromosome is equally strange: there is evidence that 150 million years ago, the X and Y were a pair of ordinary autosomal chromosomes. This means our ancestors got one copy from each parent. To convert a developing embryo into a male, the most important gene is the one aptly denominated sex-determining region Y, or SRY. The SRY evolved on one of these two ancestral chromosomes, making it into a proto-Y. This new sexual chromosome restricted its functions to testis determination through a process of accelerated degeneration during which it lost ten active genes per million years, ending up with just twenty-seven of the original one thousand genes.[13] This process captured the attention of the scientific community. A while back, some scientists argued that the Y chromosome was doomed and at the current rate will eventually disappear in eleven million years, while others noted that it has not lost any genes since human and chimpanzee lineages diverged around six to seven million years ago. Interestingly, there was a bias in the proportion of those favoring one scenario over another: more female than male scientists held the view that the Y chromosome will disappear for good at some point in the future.[14]

Due to the uniparental mode of inheritance and the scarcity of genes—meaning that no form of natural selection was expected to be strongly operating—the Y chromosome was converted into the male counterpart of mitochondrial DNA, thus reconstructing human history. It was widely used by population geneticists during the last decade of the twentieth century to build a tree of different Y lineages, following a pattern of diversifying branches. Some of the limitations of uniparental markers became obvious with the sequencing of whole genomes; it was found that past migrations followed by sex-biased admixing processes could be misinterpreted with just the observation of uniparental markers. Male-dominated migrations modified the Y chromosome landscape while affecting only about half of their resulting genomes after incoming males admixed with local women. An interesting observation is that paternal chromosomes are the most affected genetic marker in past migration episodes driven by social inequality.[15]

Disorders of Sex Development and Sex Mosaics

Exceptions to the simple chromosomal rules that determine men (XY) and women (XX) also occur. In some cases, sex chromosomes determine one sex but the person's gonads (ovaries or testes) or phenotypes indicate otherwise. This can happen, for instance, when there is a mixing of cells with XX chromosomes and XY chromosomes within the same body. Such people were previously classified as "intersex," a term that may potentially be perceived as pejorative, and their condition now comes under disorders of sex development (or DSDs), as defined in the Chicago Consensus 2006.[16] Anne Fausto-Sterling, a US sexologist who has worked on sexual identity, maintains that "intersex" individuals represent around 1.7 percent of all births, although she included chromosomal variations such as the aforementioned Klinefelter syndrome, not on the Chicago list of DSDs, in this figure.[17] A more restricted definition of DSDs indicates a prevalence of only 0.018 percent of all births.[18]

There are other instances where genitalia and sexual secondary traits do not correspond to the predicted biological sex. For example, deletions or mutations at the SRY gene cause Swyer syndrome, which is present in about one in eighty-thousand people. People with Swyer syndrome have the typical chromosomal pattern found in boys and men—XY—and female reproductive structures. Alternatively, if genetic recombination—a process in which DNA is broken down and the pieces recombined from one chromosome to another—transfers the SRY from the Y to the X chromosome, a person can have two X chromosomes and possess male traits.

Back in the 1990s, children affected by some of these conditions were assigned one sex or another, usually following their parents' difficult decisions or doctor's suggestions, and underwent surgery. This type of surgery is now controversial, not only because of the radical consequences on people's identities but also because they are sometimes performed on babies who are too young to consent to it. Moreover, as it was easier for surgeons to produce a vagina than a penis, most cases consisted of surgically turning boys into girls and telling their parents to raise them as such. As they grew up, however, as many as eight of a sample of fourteen were convinced they were males.[19] To complicate matters further, each and every person steps into a particular social and cultural environment with established religious and moral values, along with ethnic and cultural considerations that affect the person involved and their family.[20] To consider all of these variables is extremely complex, and denying the nature of biological sex and conflating it with gender further complicates the debate.

Genes in sexual chromosomes aren't the only sex determinants. Even with the onset of the hormonal process at puberty, the cell machinery that responds to these signals can respond differently. For instance, there is a situation, termed androgen insensitivity syndrome (as defined in MedlinePlus from the National Institutes of Health in the United States), that takes place when a person's cells do not respond to testosterone. As a result, people with this syndrome have Y chromosomes and internal testes, while their external genitalia display feminine traits and they develop as women during puberty (although they are infertile).[21]

Things get still murkier when scientists have been able to look at individual cells and tissues. The resulting evidence indicates that some people are mosaics: they have a patchwork of different genetic compositions due to the random processes that occur during early development. For instance, a developing XY embryo can subsequently lose the Y chromosome in some cells, resulting in whole sections of the body that are "feminine." Or two embryos can fuse together during the early developmental stages; they can remain undetected if both embryos have the same genetic sex (sometimes with strange consequences such as a mother failing the genetic test matching of two biological children).[22] Sometimes, maternal cells can cross the placenta barrier and fuse with the developing embryo.[23] Ongoing genomic projects such as the Human Cell Atlas aim to study the 37.2 trillion cells in the human body, from development to old age, to understand which genes are switched on in each individual cell.[24] We can only guess what novel findings on mosaicism await us in this new and incredibly detailed human map.

Diversity in individual biology is not exclusive to humans. A recently reported zebra finch bird, for example, appeared to be male on its right side, displaying the typical male feather pattern of an orange patch on its cheek along with black-and-white stripes on its neck, but female on the left side of its body.[25] The bird was able to copulate with females and was attacked by other males. Researchers discovered that cells on one half of the brain were genetically male while cells from the other half were genetically female. It can be argued that such cases don't invalidate the general rule of sex as being a fundamentally binary element—after all, there is not a third type of gametes. It simply tells us there are exceptions that complicate any definition, as in many other supposedly solid biological concepts, including, as we'll see in chapter 7, the species.

The Diversity in Sexual and Affective Orientation

To further complicate the question of sexual identity—crucial as it is to the individual's personal identity—biological sex is not the same as sexual or affective orientation (whether a person tends to be attracted to men or women, whatever their orientation), and it is also different from gender identity. Scientific research on these issues is rather limited, perhaps because of the possibility of misconceptions and public criticism.

It would be an oversimplification to divide sexual orientation into just two categories, traditionally labeled as "heterosexual"/"homosexual" (now an outdated way to refer to sexual orientation). The evidence indicates there exists a gradient of behaviors between extremes that cannot be objectively partitioned. Alfred Kinsey (1894–1956), in his seminal 1948 work on human sexual orientation, created a seven-category scale with data from over five thousand male sexual histories, ranging from zero ("exclusively heterosexual") to six ("exclusively homosexual"); intermediate categories included, for instance, "predominantly heterosexual, but more than incidentally homosexual" (number two) or "equally heterosexual and homosexual" (number three). Incidentally, the term "homosexual" was coined by Karl Kertbeny in 1869. Before that, it did not exist as a category of personal identity because to some extent a word creates its own reality.[26] Kinsey reported an unexpected diversity in sexual attitudes, pointing out, "The living world is a continuum in each and every one of its aspects."[27]

According to Kinsey, about 10 percent of males were more or less exclusively gay (that is, categories five and six on his scale) during at least three years of their lives, between age sixteen and fifty-five. In a subsequent study on women's sexual orientation published in 1953, similar figures were found. (Interestingly, it was only after the second book that Kinsey came under attack by conservative groups, to the point that there was a congressional investigation on his financial support and subsequently the Rockefeller Foundation ended its funding).[28] Later studies in the United States and western Europe reported an incidence of between 3 and 7.1 percent of same-sex orientation in men, and between 1 and 3.8 percent in women.[29] Moreover, the Kinsey scale is likely an oversimplification of the diversity observed in human sexual and affective orientation. Although the Kinsey report and other similar works have been criticized for potential sampling biases that may have influenced the incidence numbers, what these studies documented in the end is the existence of a large gap between social norms and sexual practices.

Interestingly, normative assumptions about affective relationships have been projected into the past, as observed in the discovery of several pairs of human skeletons in different archaeological contexts that have been dubbed "lovers." The Lovers of Modena, excavated in 2009 in present-day Modena (Italy), were dated to the fourth to sixth century CE. In what is undoubtedly an uncommon funerary practice, the hands of the two skeletons appeared to be intentionally interlocked. Viewed from the perspective of contemporary assumptions, they were described as a heteronormative couple and promoted as such in the press. Recent analysis by mass spectrometry of an enamel protein that differs between males and females, however, revealed they were in fact both biologically male.[30] The authors interpreted the burial as a unique representation of commitment between two men during late antiquity, suggesting they might have been war comrades or friends, or maybe even relatives. They could still have been lovers, although it seems unlikely that contemporaneous religious attitudes toward same-sex relations would allow such public displays of affection. Other similar examples in the archaeological record, such as the Lovers of Valdaro or the Embracing Skeletons of Alepotrypa, seem to be heterosexual couples, though the projection of present-day sociosexual ideas to the past could involve similar bias. Usually we see what we expect to see.

Same-gender and same-sex orientation is not exclusive to humans. It has been reported in over fifteen hundred animal species, both invertebrates and vertebrates such as fish, amphibians, reptiles, birds, and mammals.[31] It is in fact so common that it cannot be called "unnatural" in any sense.[32] A recent study found that same-sex sexual behavior is not randomly distributed across mammal lineages; it is especially abundant in nonhuman primates, from lemurs to apes. From an evolutionary point of view, such behavior could have been favored to mitigate conflicts in social groups.[33] As it is so prevalent, a pertinent question to ask is if such behaviors are innate (because if they are innate it means they will continue to exist in the future, irrespective of cultural circumstances). Reproductive strategies are a crucial aspect of life in sexual organisms and therefore it is plausible to assume that they could to some extent be genetically determined. Could this be the case with human affective and sexual orientation?

Some evidence, such as from studies of twins, suggests that affective and sexual orientation could have a genetic component. The underlying rationale for the twin-based approach, widely used in genetics, is that if character has a strong hereditary basis, we should expect traits to be more highly correlated in identical twin pairs than in nonidentical twins or siblings, given that the two latter groups share only 50 percent of their genes.

There have been some scientific studies that attempted to reveal biological and genetic differences between heterosexual and homosexual individuals (while referring to scientific works, I will use the original terms, even if outdated, because it is difficult sometimes to find equivalent concepts in a subject with constantly evolving terms). Yet human behavioral genetics, and in particular genetics of human affective and sexual behavior, is a controversial field prone to unreproducible results.

A study by geneticist Dean H. Hamer and collaborators published in *Science* in 1993 identified at least one type of male same-sex behavior linked to genetic markers in a specific region of the X chromosome. They reported that same-sex behavior was transmitted in families, following a maternal line (sometimes a maternal uncle was also gay). By analyzing forty pairs of brothers with same-sex behavior, Dean and colleagues were able to trace this conduct in thirty-three to a relatively large chromosomal region called Xq28.[34] The results were widely reported in the news and journalists dubbed the finding the "gay gene"; the results have not been subsequently replicated by other research groups.[35] That an issue as complex as affective and sexual orientation could be restricted to a single gene—or a single genomic region—seems unreasonable. In fact, with the passage of time, it seems incredible that a study with such a limited sample size and without data from a control group of nongay siblings was ever published.[36]

Recent studies on whole-genome-scale and large datasets have uncovered genetic variants that could explain same-sex sexual behavior. In the largest of such studies to date, published in *Science*, genetic information

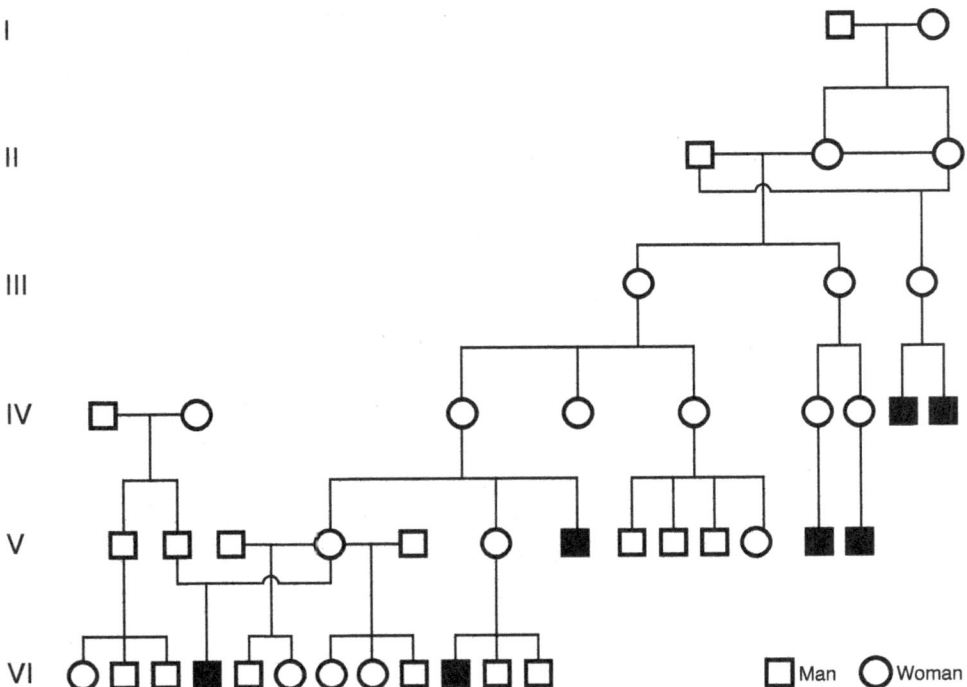

Figure 4.2
One of the family trees presented in D. Hamer et al., "A Linkage between DNA Markers on the X Chromosome and Male Sexual Orientation," *Science* 261, no. 5119 (1993): 323. Black squares correspond to gay males. Subsequent studies have not been able to replicate the original results.

from 493,001 participants from the United States, United Kingdom, and Sweden was correlated with self-reported information on sexual orientation.[37] The researchers found some markers near certain genes involved in sex-associated hormone synthesis and regulation, and also olfaction, that seemed to account for 8–25 percent (a significant but small percentage) of the variation in male and female same-sex behavior. (The authors were using a chromosomal-based definition of sex.) Nevertheless, the genes that differed between heterosexual and same-sex sexual behavior were not the same ones involved in occasional versus regular same-sex behavior, suggesting other genetic as well as environmental factors might be operating. The researchers reported that personality traits such as loneliness and openness to new experiences as well as risky behaviors and some emotional disabilities seem to be correlated with same-sex orientation too. In any case, this study shows evidence of a correlation, not a causation; a gene can be involved in many different functions that may secondarily impact a particular trait in association with certain environmental factors. To understand what it really means in biological terms, further research will be needed. Some lines of evidence point to yet another level of complexity potentially playing a role: chemical modifications to DNA (called epigenetics) that can influence gene functions at critical moments of sexual development.[38] Such modifications can be influenced by environmental factors and inherited over at least a few generations. Such a mechanism would involve a biological role in sexual orientation, but not a directly genetic one.

Fluid Gender Identities

Besides biological sex and sexual orientation, there is yet another level of complexity: gender identity. The World Health Organization defines gender identity as a "person's deeply felt, internal and individual experience of gender, which may or may not correspond to the person's physiology or designated sex at birth."[39] Gender identity can thus have more identities than woman (or girl) and man (or boy), and it can go from cisgender and transgender people to those who identify as neither male nor female, along with myriad other concepts. Although gender may well cover a spectrum, there remains psychological resistance to changing binary systems of gender traditionally viewed to derive from biological sex.

Biology alone may not suffice to understand gender identity, intertwined with culture as it is, and is inoperable in many cases.[40] The existence of gender among animals is likewise controversial because it is generally considered a human social construct (although this assumption has been recently challenged, at least in the case of our closest living relatives, chimpanzees and bonobos).[41] Two fundamental aspects in gender identity are the previous existence of socially determined gender roles—that is, the range of behaviors that are perceived appropriate for a person based on their social gender and designated sex—and the internal sense of a person's own gender identity. Going back to the Viking woman of Birka, we may suspect that certain characteristics attributed to her are ones we associate with masculine gender roles—for instance, aggressive and dominant stereotypes expected for male warriors in those times. This also illustrates the simplification of the information provided by what the biological sex represents. What is really interesting about the Birka woman is how she contemplated herself and how the rest saw her.

Nowadays it is unclear if gender is to some extent genetically or biologically based. Although it might largely be a social construct, it seems plausible that gender roles are in some measure influenced by biology.[42] For example, a large study in children's toy preferences showed that in kids between one and eight years old, boys played with male-type toys significantly more than girls did, and vice versa.[43] Despite warnings about how biased and limited previous studies on gender roles have been, these authors at least selected only those studies that were observational of children in free play and excluded those with self-reported information from the kids or their parents.

Whatever the assigned gender roles in our society are, some people perceive a mismatch between their gender identity and assigned sex. Gender incongruence is defined in the eleventh revision of the International Statistical Classification of Diseases and Related Health Problems as a "condition" (and hence explicitly not a disorder) characterized by "a marked and persistent incongruence between an individual's experienced gender and the assigned sex."[44] When this gender nonconformity causes distress, it is referred as gender dysphoria.[45] It has been estimated that approximately 0.5–1.4 percent of natal males and 0.2–0.3 percent of natal females meet the definition.[46] The fifth edition of the Diagnostic and Statistical Manual of Mental Disorders of the American Psychiatric Association gives much lower estimates of about 0.005–0.014 percent of people assigned male at birth and 0.002–0.003 percent of people assigned female.[47]

What is the origin of this condition? There have been some studies on gender incongruence in twins. A review of the scientific literature reported that of twenty-three identical twin pairs (both female and male), nine of them (that is, 39.1 percent) were concordant for gender incongruence—a significant value.[48] Nevertheless, these studies have been questioned, not only because of the small sample size but also because of the difficulties in separating common genetic background from shared environmental conditions, including psychological variables. In fact, another recent study on gender identity and gender dysphoria among a large cohort of siblings and twins in Sweden found no clear evidence for a heritable component.[49]

In a widely cited study of 1995, Dutch neurobiologist Dick Swaab reported structural differences in a certain area of the brain between heterosexual males and females, and a sample of twelve male-to-female transsexual individuals (the latter's brains being more similar to female brains, whether or not they were taking sex hormones in adulthood).[50] Subsequent studies reported that transgender men seem to have a weakened connection between brain areas that process the perception of self and body but found no evidence of feminized brains in man-to-woman transsexual individuals.[51] Recent research suggests that gender dysphoria can derive from a discordance between sex-associated hormones and brain development that can have biological and perhaps genetic roots, while some studies proposed a list of genes with uncommon variants involved in the development of a sexually dimorphic brain that could contribute to gender dysphoria.[52] With the current knowledge, it seems plausible that gender identity could be a multifactorial trait with a certain heritable component based on the additive effect of many genetic variants.[53] Nevertheless, it is likely this potential genetic component will only explain a small fraction of the resulting diversity.

Although gender might not be prevalent or exist in animals, the natural world does exhibit a whole array of diversity in behaviors related to the expression of a given sex.[54] In fishes like the peacock blenny, for instance, some males look like females to gain protection from more aggressive males and also to mate with females while other males are busy competing against each other.[55] Another example was reported in hooded warblers, in which male birds with unbanded plumage paired with color-banded males and exhibited female behavior, such as adding grass fibers to the nest and incubating. It seems to be a case of a heterosexual male accepting another male who exhibits typically female behavior and physical traits as a mate.[56]

People can argue that these are exceptions to otherwise common behaviors. Yet they've been recorded not only in captive individuals but also in wild populations under different ecological conditions. The interesting point is to ascertain that they exist at all because we tend to frame what happens in the natural world as representative of the "natural order." And the natural world is not really that well known yet. Recently, we staged an exhibit at the Natural Sciences Museum of Barcelona to illustrate different examples in nature, mainly in mammals and birds, where the observed sexual behavior does not conform to the heteronormative view of biological sex. Most people attending the exhibit, especially the elderly, were surprised to learn this circumstance is not exclusive to humans. If anything, this reveals our ignorance of the complexities of the natural world and the inherent difficulties in understanding the sexual behaviors of animals from our anthropocentric perspective.

Interestingly, as gender becomes increasingly fluid and self-determined, other identity constructs such as race are still perceived as immutable. Even if both gender and racial identification are considered voluntary constructs that are sometimes clearly linked, transracialism seems to be excluded from current social trends.[57] Such a debate raged in the United States with the case of Rachel Dolezal, a woman born to white parents of northern and central European ancestry who controversially presented herself as a Black woman.[58]

An understanding of the complexities related to sex biology transcends the anecdotal level, as it can have profound implications for people's lives and affect the civil liberties and health status of a vulnerable social collective.[59] People with sex development disorders, for example, can have fertility issues, and sometimes are at greater risk for breast, ovarian, or testicular cancers. Some authors in favor of a more inclusive agenda believe that restricting human sex to binary categories is a simplification, or worse, a justification for legal and societal misogyny and inequity, enabling the propagation of transphobia and racism.[60] Naturally, scientists, as others, are susceptible to the codes of their own ideological backgrounds, and even those who categorically deny biological sex as a binary trait can be subject to their own prejudices, well intentioned as they may be. More scientific research in these aspects of human nature is clearly the most suitable course of action.

Although in most countries of the world it is legally impossible to be anything but female or male, it is legitimate for people to feel that their personal identity is too narrowly defined by their biological sex alone. Human affective, sexual, and gender identities comprise a complex mix of biological and behavioral issues that go well beyond one's primary biological sex. Gender identity poses yet more scientific challenges due to the obvious complexity of being inserted in a specific social environment that defies simplistic classification. This is not meant to underplay the fact that subjectively defined categories might be reached by social consensus for a number of practical reasons, including, for instance, fairness in female sports. Indeed, potential biological influences in affective and sexual orientation as well as gender identity deserve to be studied scientifically, in a multidisciplinary effort, instead of being dismissed for reasons of ideological narratives—an attitude that can be socially harmful in the long run. As with the other aspects of identity explored in this book, science can have a positive impact, both at the individual and collective levels, and its dissemination can have a benign effect in countering discrimination and information. Still, beyond the basic attribution of biological sex, I suspect genetics will not be able to define or determine our affective, sexual, or gender identity.

5 Genealogical Genetics

Who might you find you have come from yourself,
if you could trace back through the centuries?
—Walt Whitman, "I Sing the Body Electric"

The descendants of the English ship *Mayflower* constitute a famous genealogical case. In 1620, a group of 102 Pilgrims and 30 crew members traveled
across the Atlantic from England to the "New World." The Pilgrims, who
were separatist members of the English State Church, wished to escape from
religious conflicts raging across Europe after the Protestant Reformation.
After landing near Cape Cod, Massachusetts, on November 21, in harsh
winter conditions, they managed to establish a settlement at Plymouth in
the lands inhabited by several tribes of the Wampanoag people. Half of the
settlers died from famine and disease during the first winter. Despite the
hardships endured, 51 of the original Pilgrims had children (in 26 families),
and today it is estimated that about 35 million people can trace their ancestry to these settlers—among them, John Adams, Marilyn Monroe, Henry
Wadsworth Longfellow, Katharine Hepburn, and Clint Eastwood. Despite
the small number of colonists, the *Mayflower* has become a foundational
myth in United States for two reasons: the first Thanksgiving celebration in
1621 (though only two foods, fowl and deer, are mentioned in Edward Win-
slow's letter describing the event) and the Mayflower Compact, a document
written and signed by 41 people on the ship focused on self-government
"for the general Good of the Colony."[1] The good of the colony apparently
did not include peaceful coexistence with Indigenous people. In 1637, the
Plymouth authorities joined an alliance of colonists against the Pequots
that ended with the killing or enslavement of about 700 of them.

A General Society of Mayflower Descendants, established in 1897, only accepts those who have documented their lineal descent from the original families. Naturally, genealogical companies are cashing in on those who would like to trace their family history back to the *Mayflower*, a not altogether trivial endeavor, though it has doubtless become a question of social prestige, and not only in the United States; there are more than 54 member societies across the world, such as the Australian Society of Mayflower Descendants. We can say the original Pilgrims have expanded well beyond their expectations.

Twelve to sixteen or more generations, however, have elapsed between the original Pilgrims and their living descendants, depending on the generational period for each pedigree. We might then ask ourselves what the probability is that today's putative descendants carry a significant genetic fraction from a specific Pilgrim ancestor. As we will see in this chapter, it is, perhaps surprisingly, close to zero.

The Family as a Root for Our Identity

Family is a clear sign of identity for us. Many consider it, in fact, the core of collective identity and the limit of individual identity. But family can extend beyond our immediate relatives—spouse, children, parents, and siblings—and it is both a cultural as well as a biological reality. One way to explore our extended family, beyond the close relatives we could come to know in person, is to search for information to build a family tree going back generations.

Amateur genealogy—research on family history—has become a popular hobby in the last few decades. A genealogy is typically constructed from census records and parish registers that keep information about marriages, baptisms, and burials. Besides the fun of doing it, the underlying psychological motivation for an ancestry researcher is that family history forms our very notions of identity and can provide our lives with a sense of purpose.[2] Along with the notion of "honoring those from the past," there is the need to understand our own cultural heritage.[3] At the same time, there is growing evidence that the pastime can unleash strong emotions, both positive and negative. Data from 775 adult Australian genealogy hobbyists indicates that nearly two-thirds of them experienced distressing emotions when uncovering, for instance, ancestors with tragic or traumatic experiences.[4]

Life stories from the past, however, are constructed in specific social and cultural contexts, and thus are continually remade because they are, in the end, narrative constructs of each generation. As Fernando Pessoa (1888–1935) says in his unfinished *Book of Disquiet*, "Because men learn only what would be of use to their already dead great-grandparents. The right way to live is something we can teach only the dead."[5] Also, we don't need to be genetically related to become someone's kin. For example, our aunt or uncle can either be a genetic or nongenetic relative (in the latter case, they would be the wife or husband of our parent's brother or sister). A family is not always understood as mainly biological or following established Euro-American concepts. For instance, in the Haudenosaunee Confederacy (called the Iroquois Confederacy by the French in colonial times), there are groups of people called clans that are considered family; each clan is linked to a common female ancestor who possessed a leadership role within it. Family names and clans are passed down from mother to child.[6] In this system, some members of your extended family are biologically related, but others aren't.

Clearly, DNA is a powerful new type of archive. Genetics can offer a new way to explore and reconstruct genealogies and family stories, among other things because it can go further back in time than currently available religious or civil registers. Yet extended families based on biological links are complex even when studied genetically. To understand our potential biological links, we need first to understand how inheritance works and builds our genomes.

Genetics of Kinship

As explained in chapter 1, you have two copies of each of your autosomal chromosomes (those that conform your genome, except for the Y and X chromosomes that determine your biological sex). You inherit a copy of each autosome from your biological mother and father. Your mother has in turn received a copy from each of her parents (that is, your maternal grandmother and maternal grandfather); the same holds for your father. Of course, you may wonder which of your maternal grandparents' copies you've inherited and the answer is simple: neither one. It's usually a chromosome formed by a section of your maternal grandmother's followed by one from your maternal grandfather, and so on, and this stochasticity continues in previous generations too. It is a mosaic, formed by the combination of two original chromosomes in a process called recombination. The same thing happens with the chromosome you pass on to your offspring: it's a mixture of fragments from your paternal and maternal copies. The frequency of recombination is not high. It has been estimated there are between two and three crossover events on each pair of human chromosomes during the formation of gametes; nevertheless, large-scale recombination affects the vast majority of chromosomes, and because of the combined impact of recombination and smaller-scale events like gene conversions and new mutations, the transmitted chromosomes are rarely, if ever, identical.[7]

. Geneticist Graham Coop from the University of California at Davis, who has been talking about these issues on his blog, put forward this metaphor: "Imagine that each of these chromosomes are books—now you could have inherited page 1–253 from your maternal grandmother and 254–600

I   II   III   IV   V   VI   VII   VIII

Figure 5.1

This scheme shows how the number of our ancestors double each generation (we are the white semicircle in the middle). At the second generation, we have four grandparents, each providing us with 25 percent of our genome, and at the third generation, the number doubles but their genetic contribution halves. Adapted from Graham Coop, "Where Did Your Genetic Ancestors Come From?," *Coop Lab* (blog), December 19, 2017, https://gcbias.org/2017/12/19/1628/.

from your maternal grandfather. In that way, the copy of the chromosomal book you receive from your mother will be a mosaic of the copies of your maternal grandfather and grandmother. The mosaic you receive was bound together carefully so that you aren't missing any pages and so you get the entire story."[8]

Thus, you inherit half of your genome from your father and half from your mother, but two generations back, it's one-quarter from each grandparent, three generations back, one-eighth from each great-grandparent, and so on. Clearly, the further back you go, the smaller the fraction contributed by each of your ancestors living at that time to your genome. But that's not all: the fraction you get from each of them is not exactly the same (except the 50 percent of our nuclear chromosomes that we get from each parent) because there are many different lengths of chromosomes and also because these do not break equally along their length; in fact, there are only 33 rupture points along our genome of 22 autosomal pairs. In practice, this means the percentages vary slightly. For instance, it can go from around 20 percent from one grandparent to nearly 30 percent from another, instead of 25 percent each. This might explain why a child sometimes looks more like one side of the family than the other. It can be demonstrated that as recently as seven generations back, some of your 128 contemporaneous ancestors likely contributed nothing at all to your genome. So it soon becomes evident that the number of genealogical ancestors greatly exceeds the number of genetic ancestors. To put it another way, a large portion of your genealogical ancestors are not represented in your DNA. We can call them "ghost ancestors."

After only ten generations, the probability of not getting a single fragment of chromosome from a specific genealogical ancestor is 57.53 percent; after a few more generations, it becomes close to 100 percent.[9] In other words, we have a 70 percent chance of sharing some genes with a fourth cousin, but the figure lowers to 10 percent for sixth cousins, which I guess explains why ancestry companies do not go beyond the level of detecting fifth cousins. At the same time, if a person wants to be sure of passing 90 percent of the genes to their offspring, they should have at least thirteen children.

Therefore even if we know we descend from King Charlemagne (748–814) by direct family links, the possibility he would have contributed something to our genome is incredibly low. In fact, Hugh Capet, the founder of the Bourbon dynasty whom we'll discuss in detail in chapter 6, did descend

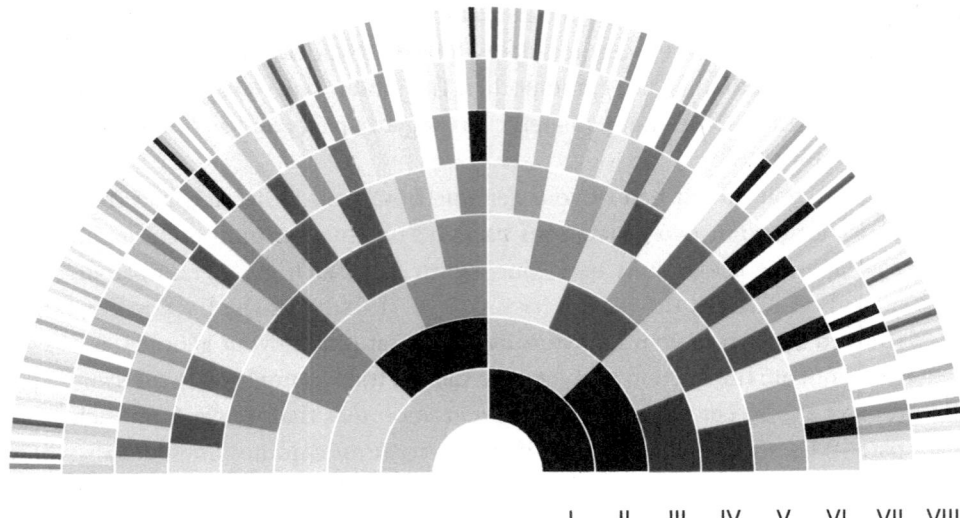

I   II   III   IV   V   VI   VII   VIII

Figure 5.2
This simulation from Graham Coop shows how much of our autosomal genome is present in each genealogical ancestor as we go back in time successive generations. Seven generations back, there are ancestors who have contributed nothing to our genome. Our genome is still there, scattered among 128 ancestors, but some of them contributed more DNA than others. As a consequence, the number of our genetic ancestors is just a fraction of our genealogical ancestors. Adapted from Graham Coop, "Where Did Your Genetic Ancestors Come From?," *Coop Lab* (blog), December 19, 2017, https://gcbias.org/2017/12/19/1628/.

from Charlemagne. At least he did through his grandmother, Beatrice of Vermandois (married to Robert I of France), who was the sister of Herbert II, Count of Vermandois, in his turn a descendant by paternal line of Pepin Carloman, son of Charlemagne. Precisely eight generations separate Hugh Capet from Charlemagne, which means he had a rather weak claim to carrying a significant fraction of his genome from the famous Carolingian king—not that this bothered him. Even in a large family, there is a small but nontrivial probability (around 2 percent) that we don't share a single genomic fragment with a member such as a third cousin (third cousins share at least one great-great-grandparent).

I have been able to reconstruct my paternal family tree back to Jaime de Lalueza, a member of the lesser nobility who married Maria Lanao in 1495 in the tiny Pyrenean village of Charo (Spain). More than five hundred years and twelve generations separate Jaime from another Lalueza, my son Marc, and it could be they share nothing of their genome. With a bit of luck, as we will see, we have at least the same Y chromosome, a rare T1a lineage, instead of the ubiquitous, Yamnaya-driven R1b.

Extrapair Paternities

Another problem that systematically emerges in genealogies is the existence of extrapair paternities. The Latin dictum "mater certa, pater semper incertus" has been updated nowadays to the humorous "mommy's baby, daddy's maybe."[10] Here again, genetics offers a new tool to study this phenomenon. A large survey of genealogical databases from the Netherlands between circa 1500 and the present day, coupled with Y chromosome genotyping, revealed low rates of extrapair paternities among the dataset, ranging from 0.4 to 5.9 percent. Interestingly, however, these rates were influenced by social and economic factors, and peaked during the Industrial Revolution among the late nineteenth-century urban working class.[11] Apparently, it is much more difficult to have an affair in a small village, where everyone knows everyone.

A recent study among Himba shepherds from Namibia revealed a surprisingly high rate of extrapair paternities of 48 percent, meaning that 70 percent of the couples had at least one child from a different father.[12] The authors point out that Himba men, rather than feeling they are being cuckolded, provide care for their nonbiological children as part of their social duties in the community. This suggests a large diversity in sexual attitudes and reproductive strategies in human societies as well as different notions of belonging to an extended family. Nevertheless, extrapair paternities, depending on their frequency, can weaken the notion of identity linked to a genealogy. Although it is quite rare nowadays (it has been recently estimated to be circa 1 percent per father-child relationship), it can happen even in royal dynasties, where bloodline legitimacy is crucial, as we will see in chapter 6.[13]

A special issue to consider in lineages of the past concerns the rules for adoption. In ancient Rome, it was common in senatorial families with no biological heir to adopt a male from another aristocratic family to secure a new paterfamilias (for that reason, it rarely applied to women), who could in turn apply for positions of power. Adoption was also a political instrument that helped to connect and stabilize upper-class families, and therefore played a role in imperial ideology.[14] Cicero, in his speeches, explains the rationale behind adoption: "What, gentlemen, is the law relating to adoption? Clearly that the adoption of children should be permissible to those who are no longer capable of begetting children, and who, when they were in their prime, put their capacity for parenthood to the test."[15] Some famous adoptions in Imperial Rome include those of Augustus, who was adopted by his great uncle, Julius Caesar, and several emperors of the Antonine dynasty (also known as the Adoptive Emperors), such as Trajan, Hadrian, Antoninus Pius, and Marcus Aurelius. Most of them, though, were related by family links; for instance, Hadrian was married to the grandniece of Trajan and Antoninus Pius was married to Hadrian's niece Faustina. It is remarkable that the first case of father-to-son biological succession in the Antonine dynasty (and indeed, the second only to that day in the Roman Empire), that of Marcus Aurelius and Commodus, proved a disaster that spelled the end of the dynasty and later the crisis of the second century.

Besides issues of legitimacy along with social and political claims in genealogies, is such randomness in genetic transmission of any consequence? Again, Coop reflects on this question:

> Does it matter that I'm not genetically related to *all* of my ancestors? . . . But any individual to whom my family tree traces back is my ancestor. My great great great great great great great great great grandmother had a profound influence on who her son (my great great great great great great great great grandfather) was, and she shaped who many of my ancestors were. Her genomic story was passed down to my grandfather and father. The fact that my father, due to the randomness of meiosis and recombination, did not pass on the small part of his genome that he had inherited from her to me seems largely irrelevant.[16]

This observation underlines the fact that all of our genealogical ancestors have shaped their descendants in one way or another. At the same time, without the existence of even one of our genealogical ancestors, we wouldn't be here, even if the lack of their specific DNA in our genomes bears no significance.

The Extended Family

An additional consequence of this phenomenon is that soon—within a few hundred years—the number of our genealogical ancestors living at a given moment will largely exceed the number of people who have ever lived on the planet, which is of course a finite—if large—amount. As Coop illustrates, going back around forty generations, by the time Charlemagne was alive (considering about thirty years per generation), you'd have a staggeringly high number of theoretical individuals in an ancestral genealogy (1,099,511,627,776); if we consider the number of potential individuals in a person's genealogy going back the forty generations, the number is of course even larger (only those back thirty-nine generations are a considerably large number of 549,755,813,888 individuals). Due to the finite number of individuals in human populations and influence of mating patterns, however, the number of unique ancestors who actually occupy the available positions in our ancestral genealogy—which we can rightly call genealogical ancestors—is much lower than the theoretical one. In addition to the number of ancestors who contributed a DNA fragment to your genome—which we can call genetic ancestors and conform a random subset of the genealogical ancestors—would be a really small number of only around 2,600 individuals.

The explanation for this apparent paradox is simple: because populations are finite, many of our ancestors are repeated and appear repeatedly in any genealogical tree, which also means that we are all in fact interconnected and share a large number of these ancestors.[17] On a larger scale—going back some tens of thousands of years—our ancestors would have lived all over the world, even in the absence of large migrations, simply through the accumulation of small movements of comparatively few people, as can be demonstrated through computer simulations.

Along with the genomic revolution, the availability of parish records coupled with current computational capabilities means we're now able to generate incredibly large genealogical databases. To take one case, researcher Gregory Clark from the University of California at Davis recently published a genealogy of 422,374 English people born between 1600 and 2022.[18] Clark used this database to explore how social status is inherited in a society that's witnessed notorious social changes over the last few centuries—including the English and Industrial Revolutions—by considering such factors as occupational status, literacy, and educational level. He found strong correlations of status, even in the case of fourth cousins, who shared (as we've seen) a tiny fraction of their genome—a consequence of having a common ancestor perhaps 150 years earlier. The correlation apparently remains unaltered across the period under consideration and concurrent major social changes.

As for inheritance of wealth, the situation is slightly different on each side of the family. The paternal grandfather's wealth is three times more accurate than the maternal grandfather's for predicting the offspring's economic status. Even if Clark shows that the patterns of inheritance of wealth are in agreement with a simple model of genetic transmission, it is likely that additional mechanisms, such as social connections based on prestige, family names, and status, are operating. Besides criticisms of Clark's ideological agenda, this work demonstrates the current computational possibilities of generating enormous genealogies.

One powerful mechanism for maintaining these differences in social status in certain extended genealogies is assortative mating—that is, the fact that marriages between individuals with similar traits do not happen at random within a population. This assortment in mating can be based on phenotypic traits such as height or social traits such as educational status that might also reflect distant family ties.[19] Therefore this process is going to increase genetic similarities between relatives, primarily for the assorted traits.[20] This and other similar studies suggest that extended genealogies reinforce their common genetic background through assortative mating, even if this sole mechanism cannot compensate for the limitations of the expected genetic transmission from a specific genealogical ancestor, as we have previously seen. Perhaps the main factor at play for being a member of a genealogy is precisely the social perception of belonging.

Interestingly, assortative mating rules aren't the only factors in maintaining genetic similarities and forming genealogies. An incredibly large genealogical study constructed from 4 million parish records from French Canadians of Quebec succeeded in demonstrating how geography had formed the genetic landscape of 1,426,749 individuals living across that region during the 400 years since the colony was established.[21] Most of the 6.5 million French speakers of present-day Quebec derive from 8,500 colonists who migrated from France during the seventeenth and eighteenth centuries. Although they settled in territories inhabited by First Nations peoples, modern genetic studies have determined that Amerindian ancestry constitutes less than 1 percent of modern French Quebecois genomes.

By combining French Canadian genetic information with that of 2,276 modern inhabitants from France, the study effectively showed that river networks in Quebec had erased the ancestral French population structure (most of the original colonists came from Aunis, Poitou, Normandy,

Perche, and Île-de-France) and formed a new, modern one. Even the circa 450-million-year-old meteor crater in the region of Charlevoix—estimated to have originally been 54 kilometers in diameter—influenced the genealogical structure across the Saguenay-Lac-Saint-Jean region to a much greater degree than the region of Provence in France. Geography and mating are continuously shaping our genomes.

How to Detect Direct Genealogical Connections

Such large genetic-genealogical studies need precise tools to screen for distant ancestry links. The main ones we have at hand are identity-by-descent (IBD) blocks. These are extensive DNA segments—up to tens of millions of nucleotides long—that are found between two people who share a recent common ancestor. The blocks are identical in their DNA sequence and delimited by breakage points determined by recombination events in our genome. For instance, it can be determined that seven generations back, our genome is present in 286 genetic blocks scattered among our 256 contemporaneous ancestors. As we have mentioned, not all of them will carry one fragment of our genome, and some of them will carry several blocks instead.

These IBD blocks are used to reveal close and relatively distant genetic relatives. Because fragmentation of chromosomes from the distant past tends to make them too short for researchers to recognize, these connections are necessarily limited to ones that occurred in the last few hundred years. Naturally, there is a correlation between the length of the IBD block and genealogical closeness; for example, second cousins will share longer IBD blocks than fourth or more distant cousins. In fact, degrees of family relationships can be estimated from the length and number of IBD blocks. This is the tool that ancestry companies use to find customers' relatives in their databases beyond the obvious level of close family.

These identical chromosomal segments can also be used to unravel isolated human groups as well as distant family links that are not easily detectable by using other population genetics methods. In a study we undertook a few years ago on modern European genomic databases, we found many IBD connections between individuals from relatively isolated populations such as Sardinian, Maltese, Icelandic, Orcadian, and Basque people.[22] Notice that while the first four are insular populations, the last likely responds to a cultural reproductive barrier—in this case, the existence of a non-Indo-European language, Euskera, coupled with certain geographic isolation. Interestingly, a subsequent analysis of three ancient Maltese Late Neolithic genomes found signs of inbreeding and small population size on this Mediterranean island.[23]

One novel use of IBD blocks involves estimating their presence in ancient genomes. A certain quality of sequence is needed for this, and obviously longer blocks are more reliable than shorter ones. Recent studies have uncovered some intriguing links between ancient genomes.[24] For instance, researchers were able to find IBD blocks in common between Yamnaya steppe nomads dated to 3000 years BCE and individuals excavated in an archaeological context of the prehistoric Afanasievo culture in the Altai Mountains of Central Asia. These individuals were family related despite being separated by several thousand kilometers—a finding that confirms the unprecedented possibilities for dispersal that the domestication of horses offered these nomads. Within the Afanasievo horizon, the researchers could identify two individuals at the level of fifth-degree relatives who were buried 1,410 kilometers apart (one in present-day Central Mongolia and the other in Southern Russia). Previous standard analyses had shown these two individuals to have had a similar genetic profile, but a precise genealogical link could only be determined through the study of these IBD blocks. As we will see, such links can also be used to connect noncontemporary people.

Linking Past and Present-Day People

The detection of IBD blocks provides sound evidence of genetic ancestors inserted in a genealogical context where the latter information was missing. In a pathbreaking study from Catoctin Furnace (Maryland, United States), the combination of IBD blocks and genomes from people from the eighteenth and nineteenth centuries enabled researchers to link individuals from the past with present-day people.[25] This archaeological context involved enslaved African people, which represents a particularly important step for a community that was ignored in records and censuses, to help African Americans reconcile with their deleted personal past.[26] Because, following the words of US abolitionist and statesperson Frederick Douglass (1818–1895), "Genealogical trees do not flourish among slaves."

Catoctin Furnace owes its historical significance to an iron forge that represents the starting point of the Industrial Revolution in the United States.

From 1776 to 1903, the rich veins near Catoctin Mountain were mined and the materials were used to produce iron implements for domestic and agricultural tasks. The furnace was active during the Revolutionary War, producing ammunition for the Continental Army, and served as the impetus for the growth of a village later populated by diverse people, including miners, woodcutters, and blacksmiths. In the early periods, a large part of the labor force consisted of enslaved Africans (at least 271 and an unknown number of free African Americans); their numbers subsequently declined by the mid-nineteenth century, when European immigrants settled the area.

The excavation of the Catoctin Furnace African American Cemetery took place in 1979–1980 due to the construction of a highway. The retrieved skeletons were transferred to the Smithsonian Institution, where they initially underwent osteological and paleopathological analysis. Recently, some of them were subjected to ancient DNA analysis that yielded workable data for 27 individuals. The researchers discovered five distinct families among the group that connected 15 individuals, while the remaining 12 were genetically unrelated. In addition, three out of 16 Catoctin males had European Y chromosomes, indicating a history of sexual dominance. In contrast, all mitochondrial DNA lineages—except one, from an individual who was clearly African European admixed and the descendant of a European woman—were from the African continent. The African connections of Catoctin African American slaves are affected by the limited number of genetically analyzed Africans. Being the most genetically diverse continent, to be able to trace the ancestral roots of tens of thousands of US customers with scientific rigor, we would need to have in the database the equivalent of literally millions of local African genomes.

The ancient genetic data from Catoctin Furnace was then compared with that of 9.3 million consenting customers from the ancestry company 23andMe, and IBD blocks were revealed between the ancient and modern datasets. It has been estimated that most of the 45 million people currently self-identified as African Americans descend from about a half-million enslaved Africans who were transported to United States from the early sixteenth century up to 1867. Amazingly, genealogical connections were found between the Catoctin individuals and 41,799 modern participants, both in Africa (mainly with people self-identified as belonging to Wolof or Congo ethnolinguistic groups) and in the United States (mostly from southern states). Also, some genetic connections were found with present-day Europeans from Great Britain and Ireland. This does not mean the Catoctin individuals are always direct ancestors of the living ones; in most cases, the genetic connections are likely through collateral relatives who descend from a common ancestor. Nevertheless, long tracks from IBD blocks were interpreted as direct genealogical links with some people. In particular, one of the Catoctin families showed direct connections to African Americans now living in the same region of Maryland, suggesting their ancestors remained there after the abolition of slavery.

The researchers explicitly stated that "our objective is to contribute to the restoration of memories of a past community whose legacy was

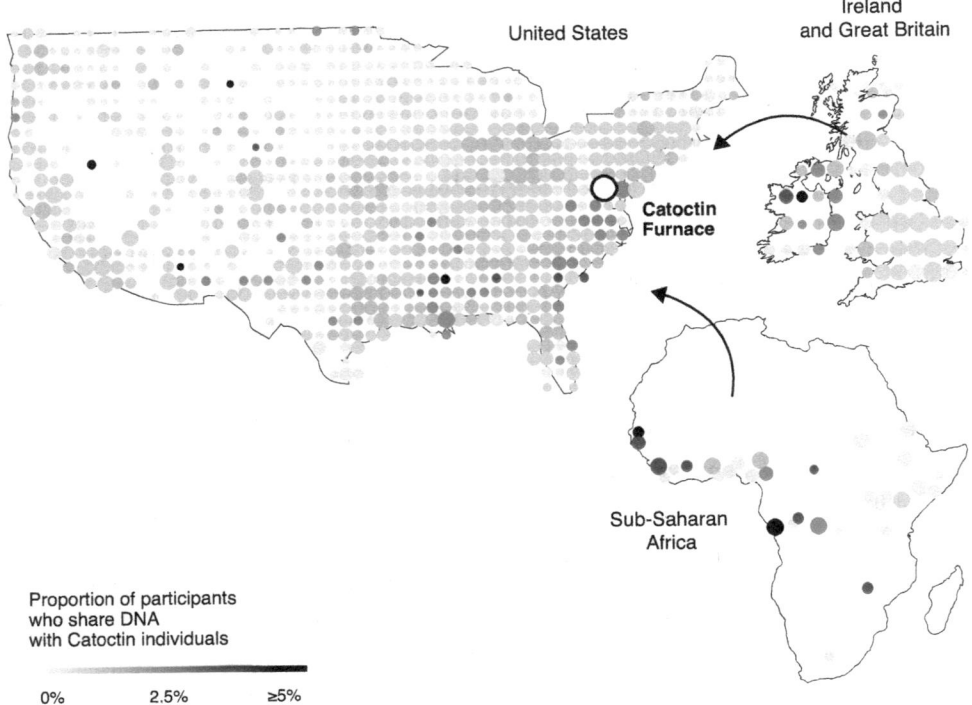

Figure 5.3
Map showing the proportion of 23andMe participants who share DNA connections to Catoctin individuals from the African American Cemetery and their connections to people in North America, Great Britain, and Africa. Modified from E. Harney et al., "The Genetic Legacy of African Americans from Catoctin Furnace," *Science* 381 (2023): eade4995.

intentionally obscured and to create an avenue for living people to learn about their ancestors."[27] Of course, a study like this also raises a number of ethical questions. To start with, the 23andMe test takers were those who consented to have their DNA data incorporated (over 80 percent of those approached).[28] But 23andMe refrains from sharing its customers' data with other researchers to protect their privacy, which means its results may not be replicated for further study by others, even though no private genetic information would be disclosed. Indeed, it seems likely that most Black American customers would appreciate that some of their anonymous ancestors in the United States have been found and that this study represents a first step toward a reparation of a long history of invisibilization and abuse. Reintegrating people into the genetic history of a country is a novel and effective way to rewrite history itself, thereby creating new identities and fostering reconciliation.[29]

While most of the ancient genomes available right now are thousands rather than hundreds of years old, soon we can expect to see more studies like the Catoctin Furnace project and thus ask ourselves what this information will mean to our sense of identity. I am a customer of an ancestry company that has already revealed hundreds of third- to fifth-degree relatives on two different continents (the Americas and Europe). No doubt it'll soon report connections to someone from the past, maybe in England (because of my English grandfather), the United States, Spain, or somewhere else. I could even be related to the *Mayflower* families; on the list of descendants from the fifth generation (which corresponds to people mainly born between 1700 and 1800), there are quite a few Fox family names—albeit Fox was not present among the original families.[30] Yet even if future genetic research finds I share a chromosomal fragment with, say, a *Mayflower* founding Pilgrim, the overwhelming remainder of my genome, perhaps 99.99 percent, will link me to other, totally different and mostly anonymous people. Will it make a difference to me? In my daily life, work, and family? Is a *Mayflower* ancestor more important than a hundred other ancestors? It could open the gates of the General Society of Mayflower Descendants for me, but what would I have in common with its actual members? Here again, the significance to one's sense of identity of a small link to someone from the past, famous or not, is unlikely to rise above the anecdotal level.

In a ceremony hold on February 17 at Catoctin Furnace, geneticists told Agnes Jackson and her daughters Sharon Green, Vicki Winston, and Barbara

Hart they were related to a toddler slave girl who died in the nineteenth century and was buried there. They shared long IBD blocks indicating the girl was Agnes's sixth- to seventh-degree relative.[31] This emotional information is just the beginning of what genealogical genetics can do. In the next few years, this emerging field will continue connecting people from the present with people from the past. For many people, this new identity layer will come as a restitution of a stolen personal history. This is just an example of the power of genetics in reconstructing, to an unprecedented level of connectivity, a temporal dimension of our extended families.

6 Family Blood: Royal Dynasties

Tout roi est un rebelle et un usurpateur. (Every king is a rebel and a usurper.)
—Louis Antoine de Saint-Just, speech, November 13, 1792

On July 2, 866, Frank commander and margrave Robert—known as Robert the Strong, I suppose, because of his imposing physique—approached Brissarthe, today a village called Pont-sur-Sarthe in modern-day Brittany (France), with his army. There he faced the combined forces of Salomon, Duke of Britanny, and Hastein, a Danish Viking chieftain. During the battle, a rumor spread that the Danish had taken refuge in a nearby church. Robert, wishing to end the battle quickly, rushed there—removing his armor to go faster—only to find it was a trap. In the ensuing melee, Robert was killed.[1] In the aftermath of the battle, Salomon was recognized as the de facto king of the Bretons by the troubled Carolingian king, Charles the Bald, until the former was assassinated in 874. The heroic life of Robert is characterized as "a second Maccabaeus" in the Annales of Fulda, an East Frankish chronicle written around the year 900.

The parentage of Robert the Strong is obscure; some historians believe his family originated in Hesbaye or perhaps from the family of Chrodegang of Metz; the name of his wife is not even mentioned in the primary sources. When he died at age thirty-five or thirty-six, he left two children, Odo (or Eudes)—who was his heir at only nine years old—and Robert. What is surprising is that the latter initiated his own royal line called the Robertians. Thus despite the difficult political circumstances and premature death of their father, both Odo and Robert ended up being kings of France. Even more remarkable, Robert the Strong was the great-grandfather of Hugh

Capet, who established perhaps the largest and oldest royal dynasty in Europe. This episode also illustrates the exceptionality of the royal houses in terms of genealogical information.

Passionate as we may be about our own genealogy, it is unlikely that most of us could go further back in time than the sixteenth century, when the generalization of parish registers was established in Catholic countries, following the Council of Trent. Our family roots are soon lost in oblivion, somehow stressing the unimportance of family background. But the royal houses are obvious exceptions. Although nowadays rare (there are around twenty monarchies in the world out of almost two hundred independent states), it was the most common political system in most countries throughout history.[2] Information on the general pedigree of these houses is available for tens of generations, and some royal dynasties represent the most extreme cases of close-kin marriages documented in humans, thereby shedding new light on the genetic definition of family.[3]

In Europe, any member belonging to these lineages, even if not currently reigning, can easily trace their ancestors several hundred years back (there are now only ten monarchies ruling in Europe: those of Great Britain, Spain, Belgium, the Netherlands, Denmark, Sweden, Norway, Liechtenstein, Luxembourg, and Monaco). The Capetian line was incredibly successful—something its founder, Hugh Capet, could hardly have guessed; after all, he had only one male heir. Hugh was elected king of the Franks after the death of the last Carolingian king, Louis V, and reigned for a short period, between the years 987 and 996. The Capetians gave rise, through multiple branches—some of them illegitimate—to thirty-six kings of France, nine of Portugal, eleven of Naples, four of Sicily, twelve of Navarra, four of Poland, two of Albania, ten of Spain, twenty of Portugal, four of Hungary, and two of Brazil as well as innumerable dukes, counts, and marquises of all European regions. From 987 to 1316, the Capetians managed to pass the throne of France from father to son over an amazingly continuous period of 329 years.

Along with the Capetians, the House of Habsburg was one of the oldest European dynasties that eventually became extinct with Charles II of Spain (another dominant royal European family, the Bourbon House, ultimately derived from the Capetians). Therefore the Capetian royal lineage is still around after more than a thousand years. As we have seen in the first chapters, the genome of Hugh Capet is literally split in thousands of small chromosomal fragments across a vast number of family descendants. French king Louis XVI, as mentioned in chapter 2, was separated from Hugh Capet by a whopping twenty-five generations along many genealogical twists, making the latter a remote ancestor of the former indeed. We have to keep in mind that death, births, and marriages within the ruling families added a dimension of unpredictability that affected genetic transmission.[4] In addition, being king or noble was in the past a dangerous job. In a study of the adult age at death of 115,650 European nobles—including 843 kings—from 800 to 1800 CE, it was found that before 1550, 30 percent of nobles died in battle.[5] This figure dropped to less than 5 percent subsequently, in parallel to a decline in violence among European nobility associated with increased meritocracy and professionalization of the armies. Hence biological factors—and chance—as well as cultural factors—and war—had more weight in the past political life than nowadays in democratic systems.

The royal families constitute hereditary lines that have tried to remain isolated from the rest of society; after all, if something becomes "mixed," it can be perceived as less "pure" and this could delegitimize the whole notion of royal exceptionalism. Thus these dynasties comprised particular combinations of genes that might not have been representative at all of the genetic variation present in the populations they were supposed to be ruling.

Something else we've seen in previous chapters is that the growing number of genealogical ancestors in a finite population means we extensively all share each other's genealogical trees to a large degree. Yet due to a tendency to marry within the family, royal lineages do this more often than commoners. For instance, an analysis of the genealogical tree connecting English king Edward III (1312–1377) to king Henry VIII (1491–1547), comprising about a thousand individuals, shows that quite a few ancestors appear more than once through different branches, and some of them up to six times in the tree.[6]

Endogamic Families and Incestuous Marriages

One obvious genetic consequence of this phenomenon can be the lack of variation triggered by reiterative marriages within the same family because of endogamic unions or, even more extreme, consanguineous unions (the later applies to marriages between relatives). The biological consequence is that mutations that might have a negative effect on an individual's survival (called deleterious mutations) can manifest in both the paternal and maternal copies of a given chromosome. This causes a phenomenon known by evolutionary biologists—especially those working on endangered species—as inbreeding depression, with potentially harmful consequences. Inbreeding depression is characterized by the reduced survival and fertility of the offspring of related individuals. In fact, it has been estimated that, on average, we all carry one deleterious mutation in our genome that if present in both copies of a given chromosome, would likely be lethal. The reason anyone reading this book remains alive is that the effect of such mutation in a gene is compensated by the unharmed copy of the same gene. Consanguineous unions would thus increase the chances of having problems with the offspring's survival. In this sense, some royal dynasties serve as real laboratories for investigating the effects of inbreeding depression on extended human pedigrees.

The negative effect of consanguinity was probably a problem in some ancient Egyptian dynasties, notably in the family of heretic pharaoh Akhenaten. Akhenaten had at least six daughters with his main wife, Nefertiti, as attested in some contemporaneous accounts: Meritaten, Meketaten, Ankhesenpaaten, Neferneferuaten Tasherit, Neferneferure, and Setepenre (the last three are only mentioned or appear in the tombs of nobles who were buried in the second half of the reign). Two other children, Meritatentasherit and Ankhesenpaaten-tasherit, are also mentioned. (As the term *tasherit* means "the minor," some Egyptologists have suggested they were daughters of Akhenaten's daughters as only he could have fathered them.) In the Amarna tomb for Meketaten, there is a scene depicting the royal couple mourning the death of a daughter and a nurse holding a high-status newborn child. A plausible explanation might be that Meketaten died during childbirth, and again, the only possible father would have been the pharaoh himself. Marriages within the royal family were frequent in ancient Egypt, especially during periods of dynastic turmoil, but father-daughter marriages (sometimes known as "royal incest") were extremely rare.[7] The only known precedent is Akhenaten's father, Amenhotep III, who married his daughter Sitamun toward the end of his reign—a strange political move since succession was already assured.[8] Egyptologist Nicholas Reeves speculates that Egyptian society would likely find such cases of intercourse deeply disturbing.[9] No doubt the king, who was convinced of his own divinity, was not someone to care what his subjects thought of his actions.

Akhenaten's example turned out to have some imitators. Ramesses II, who had an incredibly long reign, married no fewer than four of his daughters: Bintanath, Meritamen, Nebettawi, and Hentmire. The dynamics of these relationships are unclear, but they were more than symbolic unions; at least one of them, Bintanath, is known to have fathered a child of the king. On the other hand, as the pharaoh's daughters, they had no chance of marriage below their position—unless of course they married their brothers, half-brothers, or own father.

In more recent times, the Ptolemaic dynasty (also known as the thirty-third dynasty), which ruled Egypt from 322 until 30 BCE (ending with the death of the famous Cleopatra VII), established incestuous marriage between siblings as a common practice. This likely had a symbolic and political meaning because although the Ptolemies adopted Egyptian customs, including the title of pharaoh, they were in fact a foreign dynasty.[10] (Its founder was Ptolemy I Soter, a Macedonian general and probable half-brother of Alexander the Great.) As such, the adherence of the Ptolemaic dynasty to these kinds of marriages could be due in part to the limited number of royal Hellenistic candidates and the possibility of highlighting their own singularity, more than a presumed Egyptian tradition.[11] As for Greece, despite the God Zeus marrying his sister Hera, incest between siblings was perceived as a scandalous perversion.[12]

The second Ptolemaic pharaoh, Ptolemy II, married his older sister Arsinoe II (from whom he had no offspring). Ptolemy III married his sister Arsinoe III, who bore a son, King Ptolemy V. Ptolemy VI married his sister Cleopatra II, Ptolemy VIII his niece Cleopatra III, and Ptolemy IX his sisters Cleopatra IV and Cleopatra Selene. In the last generations of the dynasty, the relationships became more and more entangled. Ptolemy X probably married his sister Cleopatra Selene (after her previous husband was killed) and later on Berenice III, who was probably the daughter of Cleopatra Selene and Ptolemy IX (thus she was both his niece and step-daughter). The famous Cleopatra VII (who was the granddaughter of Berenice III and daughter of Ptolemy XII) was successively married to her brothers Ptolemy XIII (until 47 BCE) and Ptolemy XIV (until 44 BCE). In fact, she never reigned alone (in her last period, she coruled with Caesarion, the only known biological son of Julius Caesar). She was separated by nine generations from the founder of the dynasty, the original Ptolemy.

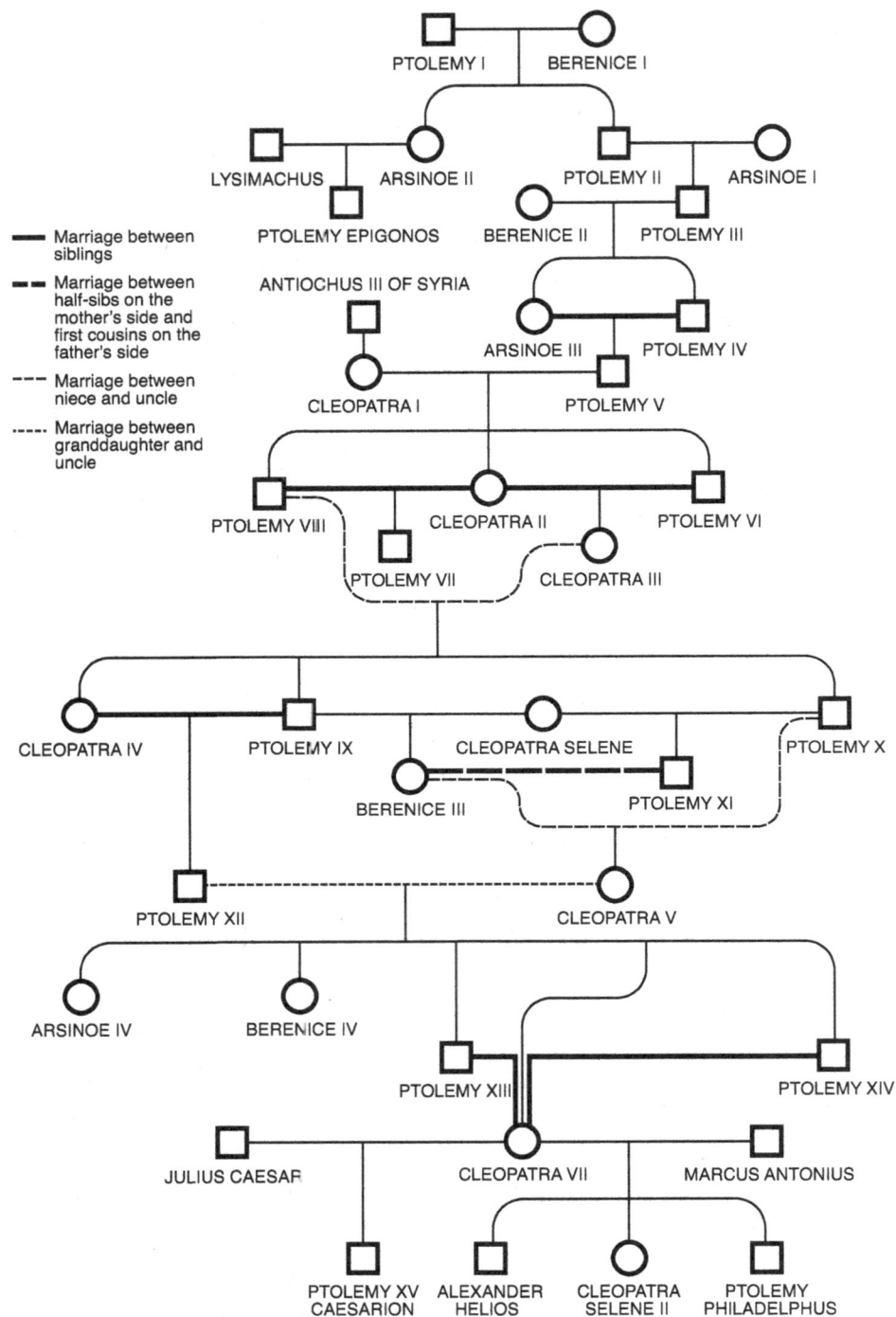

Figure 6.1
Simplified Ptolemaic dynasty family tree in Egypt. There are many instances of endogamous and even incestuous marriages (even the most famous Cleopatra—the VIIth—married two of her brothers).

It has been suggested that the Ptolemies suffered from the genetic consequences of successive generations of inbreeding; for instance, some members are described by contemporary historian Athenaeos as obese, and up
to four of the kings are portrayed as having had daytime somnolence.[13]
It seems difficult, however, to argue for the negative, cumulative effect of
consanguinity in the Ptolemaic dynasty if we look at its last ruler, Cleopatra VII. Clearly an intelligent politician and fascinating woman, she was
charming enough to make two of the most prominent Roman generals,
Julius Caesar and Marcus Antony, fall in love with her—or at least have children with her. And according to Plutarch, she could speak several languages
other than the Macedonian dialect, although certainly not Latin. Whatever the real personality and historical importance of the Egyptian queen, I
don't think anyone would contend that Cleopatra VII was impaired in any
sense as a consequence of previous generations of incest in her otherwise
complex and violent family.

Incestuous marriages did not occur in historical European royal courts,
but nevertheless consanguineous unions were not rare. The consanguinity
accumulated over generations in the Habsburg dynasty (known in Spain as
the Austrias) is well known. Researchers did a survey of more than four thousand individuals from this dynasty, spanning about ten generations over a
period of three hundred years (from 1450 to 1750), and found that 40 percent of the Habsburg marriages had a kinship coefficient higher than that
corresponding to first-cousin unions and almost 20 percent of the marriages
had a coefficient higher than that of uncle-niece unions.[14] These values
were much higher in the Spanish branch of the dynasty than among Holy
Roman emperors. Among the latter, the emperor with the highest inbreeding was Leopold I (1640–1705) and the second highest was Ferdinand II
(1578–1637). In the Austrian branch of the family, the highest inbreeding
coefficient (higher than the progeny of an incestuous union) corresponded
to Marie Antoine of Habsburg, who was the daughter of Emperor Leopold
I and his niece Margaret of Spain. It could also be estimated that Habsburg
inbreeding had an effect on both infant and child survival, accounting for
a decrease of about 13.5 percent in the probability of survival from birth to
ten years. (An advantage of dealing with a royal line is that environmental
variation—associated, for instance, with fluctuating resources—has to be
lower than that of the general, contemporaneous population.)

The general effect detected among this royal dynasty, though, is not that different from mortality found in other endogamic human populations or families. For example, the analysis of twenty-five Darwin/Wedgwood family members—which included Charles Darwin, who married his first cousin Emma Wedgwood—revealed a decrease of 5.4–9.5 percent in the probability of survival until age ten in their offspring. These findings suggest that in fact some deleterious alleles were actively purged by natural selection in the Habsburgs, thus smoothing part of the expected negative effects of inbreeding depression. The same effect has been described in certain highly endangered species such as the Iberian lynx, where a less deleterious burden than expected for such a small population has been found by sequencing their genomes.[15]

The Austrias' long history of endogamy culminated in the dramatic extinction of the Spanish Habsburg lineage, exemplified by the last king, Charles II (1661–1700). It has been estimated that he had a consanguinity coefficient—a number that equals the percentage of his genome in homozygosis or with no genetic variation whatsoever—of twenty-five.[16] This astonishingly high number is the equivalent of an offspring of two first-degree relatives, such as brother and sister or father/mother and daughter/son. Of course, this was not the case with Charles II, but he reached this record by the accumulation of ancestors marrying within the same family. Indeed, after 1550, no one from the Spanish Habsburgs married any outsider of the family. This enables us to understand why 95.3 percent of his genes could be traced back to just five ancestors. Checking his immediate ancestors makes it easier to understand the genetic mess that the royal family represented. The father of Charles II, Philip IV, was the uncle of his mother, Marianne of Austria; his great-grandfather, Phillip II, was also the uncle of his great-grandmother, Anna of Austria; and his grandmother, Mary Anna of Austria, was simultaneously his aunt because she was the sister of Philip IV. To summarize, instead of having thirty-two great-great-great-grandparents, like anyone else with no consanguineous recent ancestors, Charles II only had fourteen!

The unfortunate Charles II had a long list of ailments from birth. He had serious trouble talking, walking, and even eating, along with frequent fevers and dizziness, problems with his eyesight, sterility, and probably a certain level of intellectual disability. The king did not learn to talk until he was four and did not walk unaided until he was eight. The report on

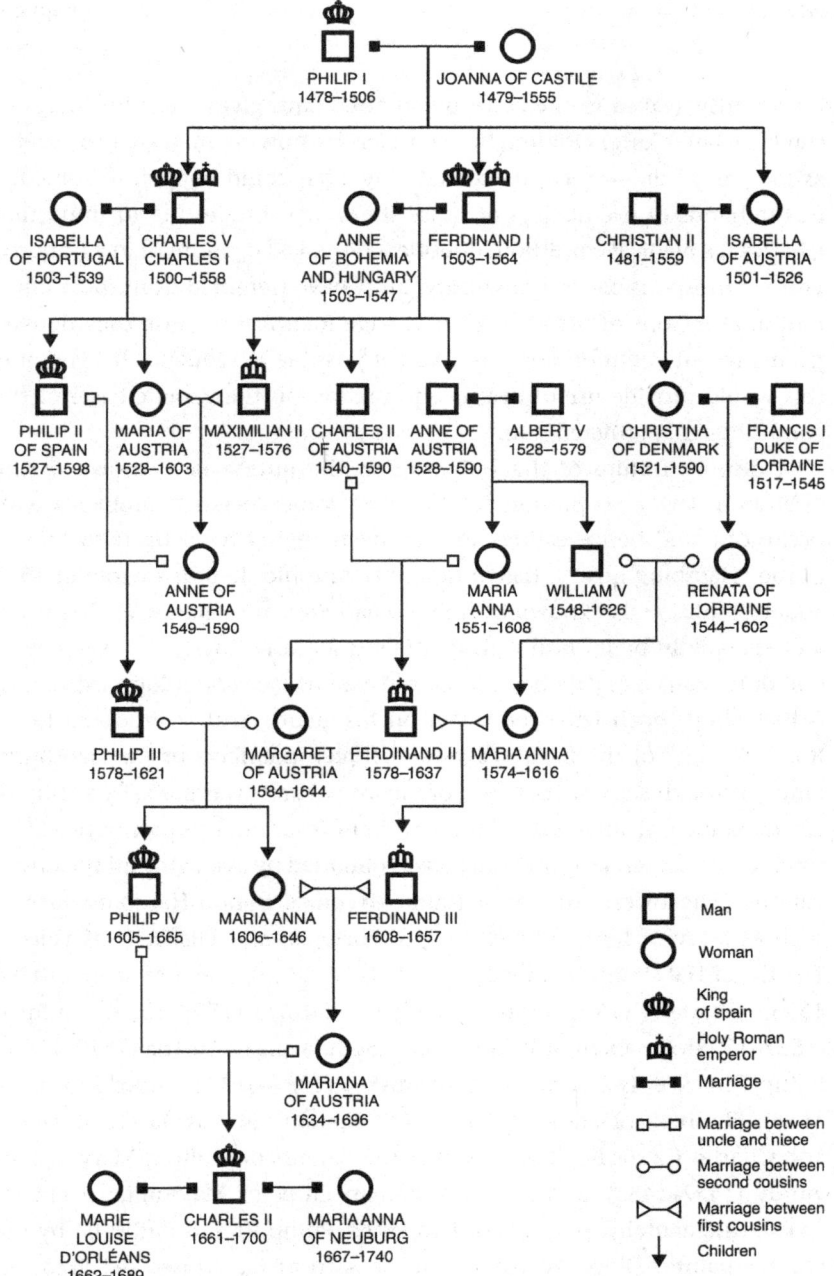

Figure 6.2
The Spanish Habsburg family tree, pointing out different levels of consanguineous marriages (marriages between first and second cousins as well as those between uncles and nieces). King Charles II was the result of a long family story of inbreeding, showing no genetic variation in about a quarter of his genome and representing the end of the dynasty.

his autopsy, leaked to the court by an attendant, gives a terrible image of the emaciated king. One might even wonder how he managed to survive as long as he did—especially considering what could happen to someone in the hands of the doctors of those times. It is interesting to think that Charles II's medical condition, doubtless provoked by more than a hundred years of irresponsible consanguinity, must have stemmed from those chromosomal regions of his genome that were identical because they derived from a recent common ancestor. Were it possible to sequence his genome, this would provide useful health information on the genetic bases of the conditions that afflicted him.

A distinct feature of the Habsburgs is the jutting jaw (known as the "Habsburg jaw"), so prominent that they sometimes had problems with occlusion—and hence eating—and an overhanging nasal tip (also known as the "Habsburg nose"). Italian diplomat Antonio de Beatis wrote in 1517 about Charles I (also known as Holy Roman Emperor Charles V), "He is tall and splendidly built, with a neat, straight leg, the finest you ever saw in one of his rank. . . . [H]e has a long, cadaverous face and a lopsided mouth (which drops open when he is not on his guard), with a drooping lower lip."[17] A study of the facial features of fifteen members of the Habsburgs family through sixty-six contemporaneous portraits was made by a team of ten maxillofacial surgeons, who rated them from mild to pronounced.[18] A total score for each king and queen was obtained by averaging all the observations. (The lowest values were found, of course, in non-Habsburg queens, such as Mary of Burgundy, Isabella of Portugal, and Elisabeth of Valois.) The list of Habsburgs included Philip I (1478–1506), Joana of Spain (1479–1555), Charles I (1500–1558), Isabella of Portugal (1503–1539), Philip II (1527–1598), Elisabeth of Valois (1546–1568), Anna of Austria (1549–1580), Philip III (1578–1621), Margaret of Austria (1584–1611), Philip IV (1605–1665), Elisabeth of France (1602–1644), Mariana of Austria (1634–1696), and Charles II (1661–1700) as well as the parents of Philip I, Mary of Burgundy (1457–1482), and the Holy Roman Emperor Maximilian I (1459–1519). (Incidentally, portraits of two kings, Philip III and Philip IV, by the famous painter Diego Velázquez can be seen at the Museo del Prado in Madrid.) The highest score obtained was, of course, for the unfortunate Charles II, followed by Philip III.

Observations from Habsburg portraits were correlated with the inbreeding coefficient (the percentage of the genome with no variation), estimated

Figure 6.3
A portrait of King Charles II of Spain nicknamed "The Bewitched" by Juan Carreño de Miranda, ca. 1685. The jutting jaw (popularly known as the "Habsburg jaw") can be observed, despite possible attempts by the painter to depict the king in a favorable light. *Source*: Wikimedia Commons.

from pedigrees of more than twenty parent-offspring generations. The researchers found a strong correlation between two features, the protruding jaw (clinically known as "mandibular prognathism") and the receding maxilla ("maxillary deficiency"), which would indicate some common genetic bases in both traits. But more interestingly, the results showed that the famous Habsburg jaw increased along with the inbreeding coefficient. These findings suggest that the genetic variants involved in this phenotype might be recessive—only manifested when the two copies are present. Yet which genes are involved remains a mystery.

This could have been an Iberian tradition. A genealogical study on the Braganza dynasty in Portugal that reigned between 1640 and 1910 found similar signs of inbreeding that could have affected the lifespan of the dynasty's progeny and decreased it by almost ten years on average.[19] The House of Braganza produced fourteen monarchs: twelve kings and two regnant queens. Consanguineous unions were common among its members; there was an uncle-niece marriage, first-cousin marriage, and several second-cousin marriages. The highest inbreeding coefficient corresponds to John VI of Portugal, who was king of Portugal and Brazil from 1816 to 1825. Another inbred member, Maria I of Portugal, who was queen from 1777 to 1816, suffered mental deterioration after 1786. She was deemed insane—as were two of her sisters—and treated in Lisbon by Francis Willis (1718–1807), the same physician who treated British king George III, although with no positive results in this case.[20]

Kings and Queens as Genetic Representatives of Their Kingdoms

Not all European royal families chose so unwisely. For instance, if we look at the eight great-grandparents of Louis XVI—the king of France who was guillotined during the French Revolution—only one of them was French (Louis, Dauphin of France [1682–1712]). Although one, Marie-Adélaïde of Savoy (1685–1712), hailed from a region traditionally placed between France and northern Italy, the rest were princesses from Germany, Poland, and Austria. This disparate choice of partners was partly influenced by politics; France was a powerful state that could turn the scales of the European equilibrium by choosing someone from another powerful kingdom. Therefore it quite often opted for rather neutral partners. For example, Louis XV married Maria Leszczyńska, a Polish princess who was daughter of the deposed king of Poland and thus had no real political power. This diversity impacts on two aspects of the French house. First, Louis XVI was a king with an extremely low level of endogamy, as opposed to Charles II of Spain. It has been estimated that less than 1 percent of his genome was identical as a result of common biological heritage—a figure that could be found in most modern, plebeian Europeans. (For instance, the previous king of Spain, Juan Carlos de Borbón, has an estimated consanguinity value of 5.22 percent, mainly because his parents were cousins.)

Second, Louis XVI's genome was not at all representative of the genetic diversity of his French subjects. His sixteen great-great-grandparents were from regions that are present-day Germany (N=eight), Austria (N=one), Poland (N=four), Italy/France (House of Savoy) (N=one), and France (N=two). Strictly speaking, only 12.5 percent of his close genealogical ancestors were French! If we could run population genetic tests on him and his contemporaneous subjects, we would find the king to be out of place among eighteenth-century French people. If anything, his genetic ancestry was quite diverse but strongly skewed toward central European regions. That may seem strange yet it is not at all uncommon. The royal British family was—through King Edward VII, the House of Saxe-Coburg, and Gotha—a German dynasty. The current name House of Windsor was adopted during World War I because of the prevalent anti-German feelings among British subjects. (Incidentally, even though inbreeding is rare among British monarchs, Queen Elizabeth II and her husband, Prince Philip, were third cousins through descendants of Queen Victoria.)

If anything, European monarchs are representative of the royal houses themselves, not of the people living in their kingdoms. If we could access the genetics of the Ptolemaic dynasty, the same situation would probably apply; alas, no burial remains of any Ptolemaic king or queen have ever been found. Doubtless they would have clustered with contemporaneous Macedonians but not Egyptians. If genetic analyses were possible, the kings and queens of Europe in the last few hundred years would generally cluster quite close to each other, like natives of a small, imaginary country, separate from the ancestry of the populations over which they reigned.

The exceptionality of these dynasties in the overall genomic context is further illustrated by a well-known episode in history: the spread of the mutation linked to hemophilia among the European royal houses in the second half of the nineteenth century and first half of the twentieth century.

Queen Victoria (1819–1901) had nine children with Prince Albert of Saxe-Coburg-Gotha (1819–1861), and about twenty of her descendants inherited this condition, making her an unquestionable carrier of the hemophilia mutation. Victoria's father, Prince Edward (1767–1820), Duke of Kent, likely had a de novo germ line mutation responsible for the condition, probably associated with his advanced age—fifty-one at her birth. Queen Victoria herself said that "this disease is not in my family," but it is obvious she disseminated it to several different European royal families—including those of Spain, Prussia, and Russia—through their daughters. (As the causative mutation is located in the X chromosome and males carry only one copy, they manifest the disease, while females can be undetected carriers.) Leopold (1853–1884), Duke of Albany, the youngest son of Queen Victoria, was a hemophiliac; he died of a cerebral hemorrhage at age thirty due to a fall. (His posthumous son, Charles Edward (1884–1954), fought on the German side in World War I and later joined the Nazi Party, before being deprived of all his titles and dying in poverty.) A daughter of Queen Victoria, Beatrice (1857–1944), transmitted hemophilia to the Spanish royal family through her daughter Victoria Eugenie (1887–1969)—married to King Alphonse XIII (1886–1941)—and another daughter, Alice (1843–1878), transmitted it to the Russian imperial family through her daughter Alexandra (known as "Alix") (1872–1918), who wedded Czar Nicholas II (1868–1918). Incidentally, the monk Rasputin's control over the hemophilia crisis of the ailing czarevitch along the disastrous path to World War I triggered the discontent of Russian society and emergence of the Russian Revolution.[21]

Many of the royal families that were in power just one hundred years ago—and had been ruling for hundreds of years in many cases—have politically vanished, with their descendants no longer ruling. We're likely to see a further dwindling of differences among the royal families as well as their respective kingdoms in this century as they mix more and more with commoners (as some of the Windsors themselves are doing).

Genetics of Legitimacy

In addition to the problems of endogamy, royal dynasties had to face questions of legitimacy—again due to the possibility of extrapair paternities as revealed by forensics. An interesting case is the recent discovery and genetic analysis of the skeletal remains of King Richard III (1452–1485), who reigned over England between 1483 and 1485 before being defeated and killed at the battle of Bosworth by the founder of the Tudor dynasty, the future king Henry VII. The life and achievements of Richard III are obscured by his fate because, after all, history is written by the victors. William Shakespeare famously offers this grim description of a villain king, equating physical with moral deformity:

I, that am curtail'd of this fair proportion,
Cheated of feature by dissembling nature,
Deformed, unfinish'd, sent before my time
Into this breathing world, scarce half made up,
And that so lamely and unfashionable
That dogs bark at me as I halt by them;

Why, I, in this weak piping time of peace,
Have no delight to pass away the time,
Unless to spy my shadow in the sun
And descant on mine own deformity:

And therefore, since I cannot prove a lover,
To entertain these fair well-spoken days,
I am determined to prove a villain
And hate the idle pleasures of these days.[22]

But was he really that bad? Armored, yielding a sword, and riding a horse to his demise (as evoked by the famous line from the same play, "my kingdom for a horse"), he seems rather unlike the kind of monster depicted here. He had a turbulent life in which he attended the coronation of his brother, King Edward IV, and fought bravely in several battles before being afflicted, it seems, by a case of scoliosis that twisted his vertebral column. On his deathbed, Edward IV nominated him lord protector and charged him with the safety of his two sons, including his presumed heir, the never-to-become Edward V. From that moment, events escalated. Richard took the sons to the Tower of London (never to be seen again and thereafter popularly known as the "Princes in the Tower"), executed some nobles loyal to his brother without trial—including the queen's brother—and propagated rumors that the imprudent marriage of Edward IV to his second wife, Elizabeth Woodville—who was not from a noble family—was illegal because he had previously married Lady Leonor Wells in secret—an assertion that only Robert Stillington, bishop of Bath, could confirm. Following this, the king declared the young princes bastards and expelled them from the line of succession. These events triggered a rebellion by part of the nobility that ended with Richard's defeat at the battle of Bosworth. The victor, Henry VII (1457–1509), revoked Richard's edict on the illegitimacy of Edward IV's descendants and married his daughter, Elisabeth of York. This marriage resulted in the union of the Houses of York and Lancaster—an alliance symbolized by the heraldic emblem of the House of Tudor, a white rose on a red rose.

What kind of burial would suit such a hated king? Chroniclers relate that the body of the king, naked and bloodstained, was conveyed by horse to the nearby city of Leicester, where it was hastily buried without ceremony in the choir of Greyfriars Church. But before long, the exact location of the burial was lost; Henry VIII (1491–1547) ordered the dissolution of Catholic monasteries during his conflict with the church of Rome, and Greyfriars was demolished. During the reurbanization of Leicester after the German bombings of the Second World War, the original place of the church was transformed into a parking lot. During the last decades of the twentieth century, academic interest in this king grew. The Richard III Society was created with the objective of improving his legacy and also to promote the search for his remains. A historian, Johan Ashdown Hill, located two living side descendants, Wendy Duldig and Michael Ibsen, nineteen to twenty-one generations down the maternal line of Anne of York (sister of Richard III) who were living in England and Canada, respectively. Thus a potential identification of skeletal remains via the matching of mitochondrial DNA was subsequently possible.

The Richard III Society, with the support of the city council, promoted the excavation of the parking lot, and the campaign started in August 2012. On the very first day, just one and a half meters below the current surface, the archaeologists found a man's legs buried there. Over the next few days, they uncovered the medieval foundations of the choir and the main nave of Greyfriars Church. Incredibly, they concluded that Richard III might have been the first skeleton they discovered at the outset of the campaign. On forensic examination, the body displayed a twisted vertebral column, compatible with scoliosis. In addition, the skull showed evidence of eight traumatic injuries made with sharp objects such as swords and halberds. (If he was in fact the king, he was not wearing his helmet when he died.) Moreover, they deduced the body was buried without a coffin, attached to the wall of a choir, his hands likely tied. Circumstantial evidence strongly pointed to the body as being the king's, though genetic data would of course be more conclusive. Turi E. King, a researcher at the University of Leicester, conducted the analysis that was subsequently published in the journal *Nature Communications*.[23]

The sequencing of the mitochondrial genome confirmed the match of the skeleton with two living descendants through the maternal line. Although the general mitochondrial lineage was a J1c2, quite common among modern Europeans, it carried a single variant not previously described in the databases. Even so, a direct match through the paternal line with the Y chromosome would be definitive. Here the researchers encountered a problem: the only male descendant of Richard, Edward of Middleham, died in infancy in 1484, and there were no registers of the potential descendants of another—illegitimate—son, John of Gloucester (1468–1491). Therefore they "climbed up" through the king's ancestors, only to go down through another branch of the family tree until a living descendant could be found. Luck was on their side; they found living descendants of Richard's great-great-grandfather, Edward III (1312–1377), through the line of John of Gaunt, first Duke of Lancaster (1340–1399). These were descendants of Henry Somerset, the fifth Duke of Beaufort (1744–1803). Of course, they were quite distant references for identifying the king, as they were separated from him by twenty-four to twenty-six generations. When the researchers compared the G2 Y chromosome lineage found in the parking lot skeleton with the modern references, something unexpected happened: none of them matched the lineage (they carried a R1b-U152 Y chromosome instead). Not only this, one of the five even had a different lineage from the rest.

Considering the ratio of extrapair paternities estimated in Europe to the significant number of generations elapsed, this result was not that unexpected. It is legitimate to argue, I think, that royal and noble families exerted strong social control over their female members, considering that legitimacy was central to their status. Nevertheless, the potential extrapair paternity detected in the Plantagenet dynasty has consequences beyond the mere anecdotal level. If this event did in fact take place between Edward III and John of Gaunt, it would mean that all monarchs descending from the latter—Henry IV, Henry V, and Henry VI—lacked the legitimacy to be kings. This would in turn affect the Tudor dynasty since Henry VII, the only king by right of conquest, based his claim to the throne partly through descent

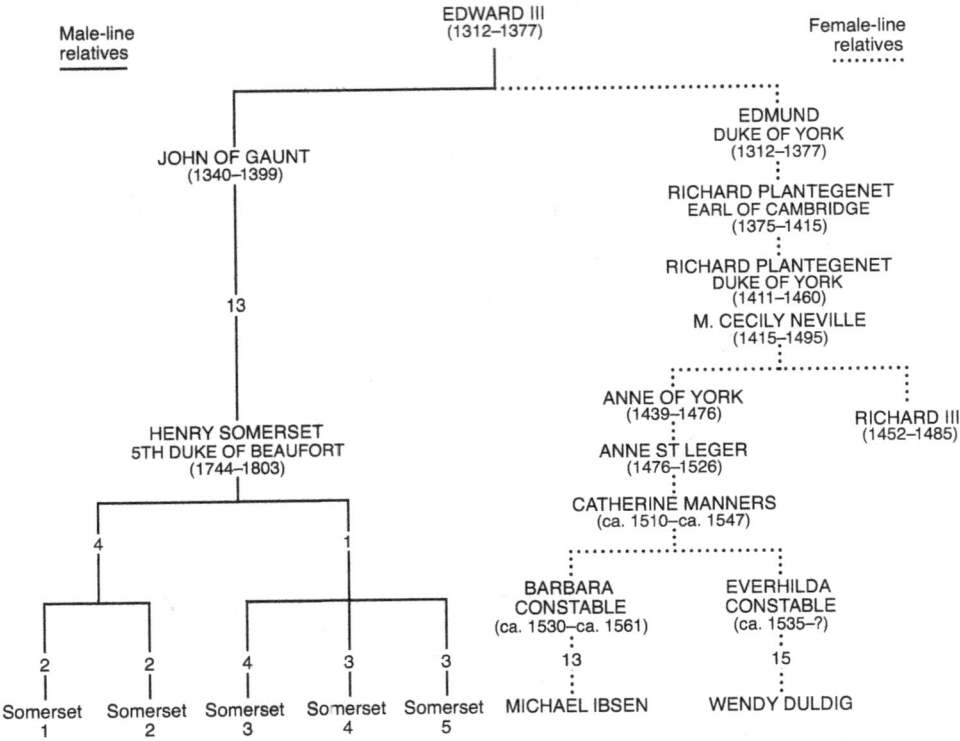

Figure 6.4

Genealogical links between Richard III and modern-day participants in the genetic study (male relatives—labeled Somerset 1–5—are anonymous). Michael Ibsen and Wendy Duldig are related to Richard III through the maternal line (mitochondrial DNA). One extrapair paternity event took place over the nineteen generations elapsed between Henry Somerset, fifth Duke of Beaufort, and Richard III. (Another relative, Somerset 3, did not match the Y chromosome lineage of the other four.)

from John of Gaunt. If the extrapair paternity took place between John of Gaunt and his son, John Beaufort, this would equally affect the legitimacy of the Tudor dynasty because the mother of Henry VII was a descendant of John Beaufort. One potential target is John of Gaunt himself; during his life, there were malicious rumors that his father was not Edward III but instead a Dutch butcher. If this was the case, the paradoxical conclusion would be that while Richard III did away with his nephews on claims of illegitimacy, his entire line, including Richard himself, were in actuality illegitimate contenders for the English throne. Some think that the issue might even compromise the current royal family. Genealogy expert Kevin Shürer, however, contended that "we're certainly not saying the House of Windsor has no legitimate claim to the throne, far from it. Royal succession doesn't work like that. There is no linear succession line between Edward III and Elizabeth II. Yes, they are related, but the whole point of monarchy is that over several centuries it takes various twists and turns. Monarchy is about opportunity and chance as much as it is about bloodline."[24]

Wealthy Bloodlines

We may not think royal families are that common anymore, but there are indeed dynasties of the wealthy, and one of the reasons for endogamy within the royal dynasties—that is, accumulation of power—remains in effect. Among such wealthy clans as the Rothschilds, a similar strategy for marriage within the family has been followed.[25] The founder of this banking family, Mayer Amschel Rothschild (1744–1812), barred female descendants from direct inheritance so they had no chance to maintain their social status unless they married other Rothschilds. It is not surprising, then, that four of Mayer's granddaughters married four of his grandsons, and one married her own uncle. Between 1824 and 1877, of thirty-six available male Rothschilds, thirty married their cousins—preferentially from different branches of the family business.[26] The Rothschilds are not an exception; the same thing happened within the DuPont (sometimes spelled du Pont) family, one of the richest in North America (though originally from France). The famous entrepreneur Pierre S. DuPont (1870–1954) married his first cousin, Alice Belin, in 1915. (They had to marry in New York because his home state of Pennsylvania prohibited intercousin marriages.) This seems to have been a family rule, as famously expressed by patriarch Pierre Samuel du Pont de Nemours (1739–1817) in a letter to his son Eleuthère Irénée on January 26, 1810: "If your boys were not younger than your girls, the marriages that I should prefer for our colony would be between the cousins. In that way we should be sure of honesty of soul and purity of blood."[27] The first consanguineous DuPont marriage took place in 1833 between Sophie Madeleine du Pont (daughter of Irénée) and her first cousin Samuel Francis Dupont (son of Irénée's brother, Victor), and despite orders from the then head of the family General Henry du Pont (1812–1889) to stop them because of a remarkable child mortality rate in the family, the practice continued to take place through most of the nineteenth and beginning of the twentieth centuries.

Those who assume that consanguineous or even incestuous royal bloodlines are exclusive to well-known, powerful kingdoms of the past such as the Egyptian, Persian, and Inca Empires will be surprised to learn of the genetic findings on skeletal remains excavated from an enormous megalithic passage tomb in Ireland, Newgrange (a UNESCO World Heritage Site). Although some archaeologists have speculated that the construction of such massive monuments, with the consequent huge expenditure of labor, was a cooperative effort based on a common ideology, an alternative and equally plausible view is that it was a reflection of a highly hierarchical society. In 2020, genetic evidence from the remains of an adult male discovered within the central section of the Newgrange tomb proved him to be the offspring of the union of first-degree relatives.[28] Long sections of the individual's chromosomes showed no variation at all (a sign of recent consanguinity, meaning that the two copies derived from parents who were related), totaling one-quarter of his genome. The data indicated the parents were first-degree relatives, but whether the union was between siblings or parent and offspring could not be determined (for biological reasons, the latter case would likely be father and daughter). As if that weren't enough, a recent fictional character from the *Game of Thrones* saga, Daenerys Targaryen, has an estimated 37.5 percent of her genome with no variation at all due to long-term inbreeding in the family.[29]

This is the only case of incest so far found in European prehistory and points to the existence of a politico-religious elite who could socially accept incestuous unions so as not to dilute their bloodline. When political leaders are considered divine, they perceive themselves as exempt from social convention; this also suggests that the rest of society did not follow the same rules, and in fact, genetic analyses of more than forty contemporaneous Neolithic remains from Ireland and other European regions do not detect a substantial level of endogamy. (A potential exception would be a son of second- or third-degree relatives in a megalithic burial from Sweden.) An interesting inference is that these early Neolithic communities, even if necessarily small in number, possessed social attributes that would be found in later, large states such as Egypt. Interestingly, an analysis of more than four hundred genomes now sequenced from European prehistory has made it possible to trace potential signs of endogamy over time. Researchers found that signs of genomic endogamy decreased over the last ten thousand years in parallel with the increase in population associated with agriculture. Although some highly inbred individuals are found in the dataset, they are rare.[30]

Though we may not belong to a royal dynasty or wealthy family, we can genetically measure the possibilities of inbreeding among our own ancestry. Aside from particularly small and isolated populations (nowadays religion, more than geography, is the main factor driving isolation), our common pedigrees extend through—and get lost in—the anonymity of history and are difficult to reconstruct. Recently, a new combination of genomics and computational analysis on genealogical databases has enabled the reconstruction of incredibly long and complex pedigrees that were until now simply impossible to contemplate.

One study made it possible to reconstruct a single pedigree encompassing thirteen million people (selected from among a database of eighty-six million people called Geni.com, painstakingly built by genealogy enthusiasts).[31] The researchers were able to go up to eleven generations back and uncover some interesting aspects on marriage trends. For instance, prior to the Industrial Revolution (before 1750), most marriages, both in Europe and North America, occurred between couples born within just ten kilometers of each other. But after 1870, with the widespread availability of various modes of transport that revolutionized human communications (first bicycles and then railroads), that distance increased continually until by 1950 it reached about a hundred kilometers on average. (This also means it is increasingly difficult to find your better half in your neighborhood.) Between 1650 and 1850, the average demographic data are compatible with family ties expected for the level of fourth cousins (meaning of course that many marriages were effectively between first cousins). In subsequent years, every increase of seventy kilometers between couples correlated with a decrease of one kinship generation degree. Interestingly, the researchers observed a fifty-year decoupling between intermarriage dispersion and the decline of genetic relatedness between couples, meaning there was some resilience in abandoning what was primarily a cultural practice. That said, consanguineous traditions—mainly marriages between first cousins—are still prevalent in some Muslim and Jewish communities.[32]

Most of us are now free of any genetic signs of endogamy or consanguinity in our genomes. Most readers can say with a high degree of probability there are no family links among their immediate ancestors, and, with an ever-growing world population and increased transport options, it is likely that the sort of intrafamily marriages seen in royal dynasties through history—as well as in some wealthy families of the Industrial Revolution— are a thing of the past.

An interesting point to consider is that dynastic legitimacy does not equal genetic descendance. As we saw in the previous chapter, several generations on, chances are really small that we carry a significant fraction of the genome of a particular dynastic founder, so the glorification of a royal ancestor is more a political than genetic reality. Royal legitimacy is indeed, as stressed by Shürer, a mixture of bloodline and opportunity. These families should probably be viewed as political as well as biological units, often subjected to a fascinating set of accidental, historical circumstances. As a consequence, as we have seen, kings and queens usually do not represent the overall genetic ancestry of their subjects. Rulers and ruled are genetically decoupled.

And contrary to our expectations, even well-known historical genealogies such as royal families can be genetically inconsistent because the existence of extrapair paternities. With enough time, it is likely that most genealogies will be interrupted through the paternal line at some point; I guess this could be contemplated as the secret females' revenge for their long history of male domination. I predict future ancient DNA studies will uncover more family secrets of illegitimate descendants, including high-status families. To offer just a recent example, the genetic analysis of two Jamestown settlers who died within weeks of each other in 1610, Sir Ferdinando Wenman (1575–1610) and Captain William West (1586–1610), suggests the latter was the illegitimate son of Elizabeth West, daughter of Thomas West, First Baron De La Warr, who died unmarried.[33]

In practical terms, illegitimacy and extrapair paternities are yet another warning against stressing the importance of our potential genetic links as a sign of our identity. Even with all the genealogical information at hand, the genetic connections can still be uncertain.

7 Collective Identities: From the Clan to the Population

Most people are other people.
—Oscar Wilde, *De Profundis*

German naturalist Johann Friedrich Blumenbach (1752–1840) had access to a geographically diverse set of 240 human skulls to study the whole range of human skeletal diversity. One of these was the dissected skull of a young Georgian woman captured during the Russo-Turkish War (1787–1792) who died in prison (this rather obscure war involved the failed attempt by the Ottoman Empire to regain the Crimean khanate from the Russian Empire). Blumenbach praised the symmetry of the Georgian skull and commented on "the admirable beauty of its formation."[1] The skull served to name Blumenbach's own race: Caucasian, for whites. This strange aesthetic quality in a single skull was going to have a long and surprising scientific persistence.

Blumenbach is widely seen nowadays as the inaugurator of physical anthropology, a discipline that seeks to study the origins and evolution of the human body. As this scientific field was embedded in the development of racial science, so is its founder. His legacy is rather controversial. Although he promoted the view that humans were divided into five main races—in his initial observations there were only four—he supported both the unity of the human species and the abolition of slavery.[2] Today, Blumenbach is better remembered for his work on the quantitative study of human skulls, where he promoted five varieties or races: Caucasian (Europeans, North Africans, Near Easterners, and South Asians), Mongolian (East Asians), Ethiopian (sub-Saharan Africans), American (Native Americans), and Malay (which included Southeast Asians and Pacific Islanders). After his death, his ideas of equality were reframed into a racist discourse that spread across the Anglo-Saxon world and Germany.

Subsequent race scientists promoted a view that the other races degenerated from the primordial Caucasian "beauty" due to environmental factors such as harsh climate, excess sun, and poor diet. Equally surprising is the fact that "Caucasian," or the more ambiguous "Caucasoid," despite its former use in racial classifications, is a label still widely used in scientific studies. Between 2010 and 2021, for instance, nearly five thousand biomedical papers still used the term "Caucasian" for European populations.[3] And not only in academia: every time I have to define myself as "Caucasian" on surveys or the census, I recall Blumenbach's Georgian skull and wonder what I have in common with that poor eighteenth-century woman.

Despite his posthumous success, Blumenbach was not the first to try to capture human diversity by ordering it into recognizable natural categories. It's a concept that somehow derives from the ancient Greek idea of the existence of Platonic types—although it is likely older. Starting at the end of the fifteenth century, the Europeans arrived in "new" lands in Asia, Africa, and the Americas, only to find they were already inhabited by other people. To justify their rights to occupy these lands, a central idea emerged: that the natives of these lands were mentally and culturally inferior and that taking control of them was in fact an act of "civilization" (coupled sometimes with the mission of spreading the "true" religion). During the Enlightenment, the application of the scientific method was used to justify political domination, and the Cartesian conception of the natural world implied that human variation could be classified by objective means.

Decades before Blumenbach, Carolus Linnaeus (1707–1778) led the effort to classify the natural world into categories. He coined the term "Homo" for our genus and "Homo sapiens" for our species. In the case of humans, however, his fundamental role in taxonomy had deleterious long-term effects. Contrary to other species, he subdivided ours into four varieties: peoples from Europe, from the Americas, from Asia, and from Africa. (He later added, rather bizarrely, two more groups: "Homo ferus"—wild children—and "Homo monstrosus"—abnormal people who, according to Linnaeus, included "Hottentots.") To define each variety, he mixed physical traits such as hair and skin color with perceived behaviors and moral attitudes. For instance, Europeans were "light, wise, inventor[s]" while Africans were "sly, sluggish, neglectful."[4]

The Racial Science

Linnaeus's classification system paved the way for subsequent scientific racism; Blumenbach's study of the skulls provided a supposedly objective scientific tool.[5] Racial prejudice—or prejudice against the "other"—might be a universal feeling that exerts a powerful influence on the perception of people belonging to a specific group, even if these groups are nonstatic, cultural fictions. Not only does it have the role of consolidating adherence to one's own "race" by identifying the "different people" but it also acts as a feedback system to justify discrimination against other races or groups. Races were understood to be biologically differentiated entities that could be defined by hereditary traits. The race theory projected certain abstract constructs, named racial types, and tried to fit all humans into these previously selected types; this is an intellectual process of reification. Dealing with preconceptions to classify human diversity, however, meant that social and personal prejudices could have a large influence on anthropological science—likely much more than in any other scientific field. In theory, race could be approached scientifically; the notion that our species could be further divided into taxonomic units could be tested. But it was impossible to test this hypothesis rigorously without the genetic level. As constructs versus real natural entities, races kept the flexibility of being named and redefined at one's convenience. The race scientist had the job of fitting the observed diversity (usually physical, but behavioral and even moral too) into previously defined categories. And once the variation had been classified, the race scientist unsurprisingly tried to order them in a hierarchical way, indirectly justifying differences in rights as well.

Later in the nineteenth century, race scientists, having erroneously digested the evolutionary ideas of Darwin, found a plausible explanation for racial scales of progress: the existence of races that were less evolved than others. Peoples of the African continent were consistently seen as less evolved and placed on the lowest branches of the human tree, which was used to justify colonialism and even slavery. Therefore racial classifications didn't remain a futile intellectual exercise. In fact, they turned out to have enormous social and political consequences in the twentieth century, culminating in the atrocities of the Nazi regime.

German anthropologist Walter Jankowsky (1890–1974) stated in 1930 that "race is a matter of fact; they exist with independence of the scientific research on races."[6] Notwithstanding such naive optimism, there was absolutely no consensus on the number or definition of these races; numbers ranged from four to more than thirty, and each author had their own classification. For example, Thomas Henry Huxley supported four in 1863, Ernst Haeckel thirty-six (in twelve species!) in 1873, Paul Topinard sixteen in 1878, Topinard again nineteen in 1885, William Flower three in 1885, Joseph Deniker thirteen in 1889, Deniker again twenty-nine in 1900, George Montandon twenty in 1929, Egon Von Eicksted four in 1934, and Carletoon Coon five in 1939. One could argue that this is how the scientific method operates after all, by creating new categories and testing them to see if they fit into the observed variation better than previous ones. Yet there were likely other mechanisms involved in this proliferation of races, such as formulating specific categories for geographic areas being subjugated by their respective colonial powers or the need by some anthropologists to establish their own academic profile.

It is perhaps amazing that the only consensus was that races existed. Even Darwin, a fervent abolitionist, held conventional views on gender differences and human diversity.[7] He was convinced about the existence of races and that their differences were not only physical but also emotional and intellectual.[8] The reason these categories were perceived as biological is that they mixed features like language, religion, and even apparel with certain physical traits to help circumscribe entities that were to some extent biological. As Jonathan Marks points out, "To understand race properly, however, we must appreciate that it is a biocultural category, the result of a negotiation between patterns of difference and perceptions of otherness."[9] Most racial categories were based on certain phenotypical traits, notably pigmentation. But such traits were misleading indicators of genetic variation because their geographic stratification is quite unusual in our species. Not only are there biological and cultural dimensions; the concept of race is embedded in subtle class constructs, such as the racial identification of sports figures in the United States. (African Americans, for instance, are overrepresented in the National Basketball Association but underrepresented in Major League Baseball.)[10]

Although race seems to have vanished from the current scientific literature, the search for collective identities persists, even in genomics. In a recent search of 11,635 published papers from the archives of the *American Journal of Human Genetics*, which has one of the longest records in the field, researchers tracking terminology used from 1949 to 2018 found that the term "race" appeared in 22 percent of the papers in the first decade, but in only 5 percent in the last one, meaning that most geneticists have effectively abandoned the use of the term.[11]

Still, we've already seen in previous chapters how genomics relates to our individual and genealogical identities. We can then ask ourselves, What about belonging to a collective? Even if races don't exist, the search for collective identities is still present in modern societies. What can twenty-first-century genetics tell us about our supraindividual identities?

Genetics and Social Organization of Small-Scale Societies

There are different levels of social organization that can define a human collective, from clan to tribe to nation. The definition of these categories is often disputed by social anthropologists. It is likely a subject that defies generalization since it is conflated with notions of cultural and linguistic identity. For instance, clans can be defined as social units of people who descend—or claim to descend—from a common ancestor, either matriarchal or patriarchal. It is larger than a family, so usually the members of the clan assume a shared ancestor even if they cannot enumerate all the family links. Clans are often small groups of no more than thirty to fifty people, though some might be larger, including up to thousands of individuals. In small-scale societies, all boys from the same generation within the same clan are called "brothers."[12]

Many clans are organized patrilineally, though some are, say, nonhierarchical or "headless" and organized in subclans or different lineages, such as the Nuer of Sudan studied by E. E. Evans-Pritchard in the 1930s.[13] These lineages regard each other as distant relatives in a broad sense, but genealogical links are only recorded within close kin units; they can function autonomously, yet they can eventually unite at a higher level if circumstances demand it. In African groups, each clan has a "mystical" connection with its land and, for instance, stages annual rites to appease its ancestors' spirits and ensure a good harvest. Thus ancestral cults could be important not only to strengthen the clan's identity and cohesion but also to ensure its survival. Even within the Nuer, though, not all members can genuinely claim descent from their original ancestors, and a significant fraction of them are in fact Dinka adopted into Nuer society.[14] As in other examples studied by social anthropologists, it seems having a common language to describe political relationships is at least as important as actual kinship.[15]

Going up a hierarchical level, a tribe is usually considered a group of people larger than a clan who are united by shared signs of identity such as culture and language. In modern anthropological literature, however, tribe is often a discredited category, partly due to the difficulty of devising a definition and partly because of its widespread former use in colonial ideology. Usually tribes are just convenient colonial constructs. That said, the term continues to be used outside the field of anthropology, mainly due to the longevity of the idea of "primitive" societies, although in some forums it has been replaced by the equally misleading term "ethnic group."

From social anthropological studies it can be concluded that most societies are both endogamous and exogamous to certain degrees. Whether they are one thing or another may well vary depending on their perceived notions of sameness and otherness at different levels (clan, cultural collective, nation, etc.). In many small-scale societies it is considered desirable to marry someone from your kin but not someone who could be perceived as close blood kin.[16] Of course, who is socially acceptable as a reproductive partner is not strictly specified and can vary across cultures.

Irrespective of their disputed nature, social units such as clans and tribes share a time dimension that in principle sciences studying the past should be able to explore. But comparisons with the present are difficult to make. Most small-scale societies are being disrupted by statehood or nationhood as well as the dynamics of globalization. It can be argued that many stateless social units are quickly fading away on a large part of the planet. Thus trying to extrapolate present-day anthropological observations about social structure to past societies from which we have no precise information on kinship or ancestry is a challenging task. In prehistoric times there were small-scale societies across Europe, first as foragers and later on as agriculturalists. During the Neolithic period, some of these societies persisted for hundreds of years and built megalithic structures still visible today that must have required the sustained effort of their communities to construct. We can ask ourselves, How were they organized as a society? And if we know this, we can subsequently ask, What were the signs of identity that glued these communities together?

But all we can obtain from the study of the skeletal remains of past people is a confusing picture of a collection of fragmentary bones, accumulated generation after generation (how many is sometimes difficult to know because some dating methods are not always that precise) at a particular archaeological site. Even were we to investigate these bones with the most sophisticated anthropological tools, only information on certain traits, including sex and age attribution, could be obtained, without any proof of the basic social and genealogical clues—that is, until recently.

Thanks to the revolution in ancient genomics, we can now study the genetic structure of past social organizations in places like the European continent, where such small-scale groups disappeared hundreds of years ago. Hence genetics can not only shed light on individuals but also explain how these individuals are related, from which social structures can be deduced. With this information, we can glean clues about social organization and likely human collective identity in the last few thousand years.

Let's examine the results from a recent work from a small-scale prehistoric society, represented by the skeletal remains of many of its members excavated at Hazleton North barrow near the village of Hazleton at the Cotswolds, in Gloucestershire (South West England). The barrow was an Early Neolithic long-chambered cairn. Located in a field under cultivation, the site was being damaged year after year by plowing, although by 1979 the top of the structure was still unplowed.[17] It was excavated between 1979 and 1982 before being dismantled. (The main chamber has been reconstructed and can be visited in the Corinium Museum in Cirencester.)

The cairn was trapezoidal and enclosed by a wall measuring fifty-three meters long by eight to nineteen meters wide, with two L-shaped roofed funerary chambers near the center of the structure that opened to opposite sides of the cairn; they are called the south and north chambers. Each of the two burial chambers were comprised of three spaces: an innermost chamber, a passage, and an entrance. Although both had lost their roofs, numerous burials were found undisturbed during the excavation, accounting for a total of twenty-two adults and at least nineteen preadult individuals. (In addition, the excavators found remains of three cremated individuals at the north entrance, corresponding to the latter stages of use.) The original excavator, Alan Saville, hypothesized that this cairn—like other similar ones in the region—was used as a burial place for only a few generations before being abandoned as funerary practices and probably farming settlements changed during the Middle Neolithic period. The findings were radiocarbon-dated to about fifty-seven hundred years ago.

In 2021, genetic researchers were able to obtain DNA from human remains at Hazleton and reconstruct the genealogical relationships among thirty-five individuals inhumed there.[18] These links can be assessed following the principle that first-degree relatives—such as parents and offspring or brothers and sisters—share 50 percent of their genes—a percentage that halves each generation (we share 25 percent of our genome with each of our four grandparents, etc.). With enough genetic data, it is even possible to distinguish between different types of first-degree relatives. This is because between parents and offspring the shared 50 percent constitutes a copy of their genomes, and therefore distributes uniformly across chromosomes, while between siblings the identical chromosomal fragments—deriving from identical chromosomes from the parents—alternate with fragments with genetic differences—deriving from the alternative chromosomes. The genetic researchers were able to reconstruct the genetic links among twenty-seven individuals, spanning up to five generations, in a large multigenerational family. (Some individual skeletons were missing from the cairn although the researchers knew, on the basis of the genetic results, they must have existed.)

The reconstructed pedigree showed a founder male—who researchers informally dubbed "The Patriarch"—who reproduced with four different females, one of them unsampled. Also belonging to this clan are male lineage descendants, their female partners, and, interestingly, a few males not linked to the main lineage. The latter presence suggests that close kinship was not the sole criteria for inclusion in the cairn's tombs. A majority of males shared the same paternal Y chromosome (an I2a1b1a1a1 lineage), indicating that the main links across generations were through the paternal line. Moreover, there was a bias toward more males than females in this cairn (twenty-six out of thirty-five). The same bias is observed in other Neolithic collective tombs in the British Isles, which suggests that deceased women could be subject to different treatment or perhaps had a lower status. Within the Hazleton cairn, there seem to have been differences between the south and north chambers. In the latter, the bones of at least five individuals showed signs of gnawing by canid scavengers and weathering, implying that they were exposed to the open air for a while prior to being interred in the chamber.

A lack of adult daughters corresponding to the main lineage (none versus fourteen sons) suggests that these women were interchanged with neighboring clans as a way to establish alliances. In addition, the mothers

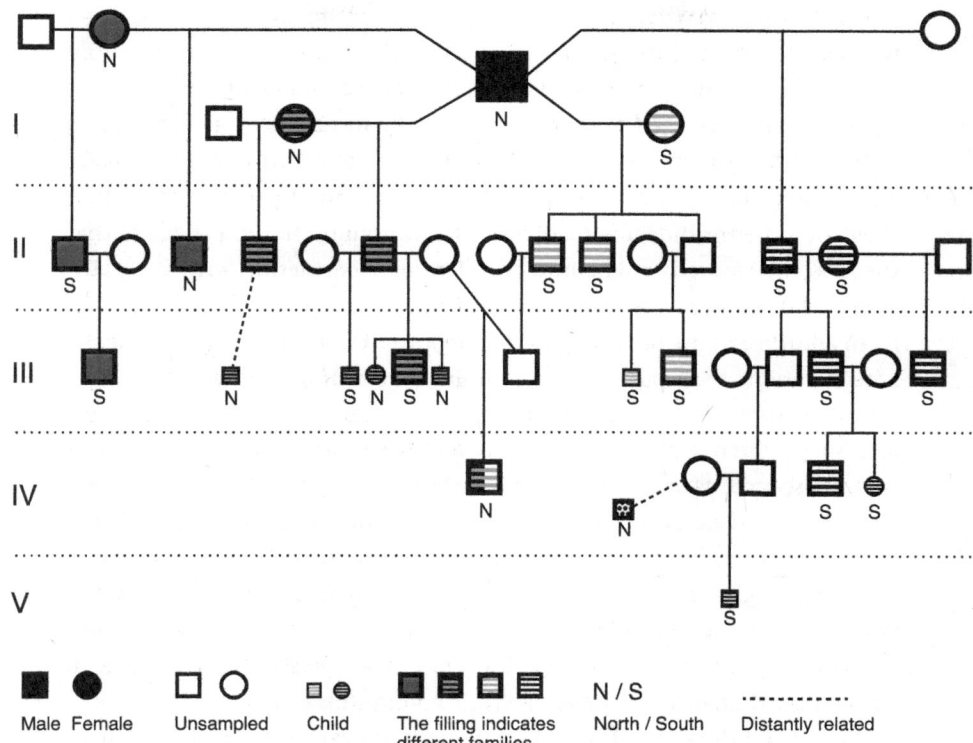

Figure 7.1
Reconstructed family tree at the Neolithic site of Hazleton (England) by using ancient DNA. S and N indicates the south and north chambers. Modified from C. Fowler et al., "A High-Resolution Picture of Kinship Practices in an Early Neolithic Tomb," *Nature* 601 (2022): 585.

of the males within the Hazleton lineage had no obvious relatives besides their biological offspring—that is, they had no parents or siblings in the group. The presence of daughters who died in childhood indicates that such female mobility did not take place until reproductive age.

The most remarkable finding is of course the presence of a patriarch who founded the clan and had children with four different women. Although it is not possible to know whether this was a case of serial monogamy, it seems unlikely considering he had at least six children (three of them with a single woman and the rest with each of the three other women). It is entirely plausible that this was a case of polygamy since another instance of

one man with two women—the man being one of the sons of the founder—
was also observed in the second generation. Keep in mind that only those
unions that had descendants buried in the cairn could be observed, so the
patriarchal pattern seen here could actually be more extreme. At least two
of the patriarch's reproductive partners also had children with other males
who were distantly related to him (the inference from genetic data is that
they could be third-degree relatives). This intriguing finding suggests that
these women were brought into the clan after partnering with the patriarch's relatives—or otherwise cuckolded the patriarch.

In addition to the obvious paternal organization in the clan, other social
considerations took place, as an examination of those buried in the two
chambers showed. All twelve individuals descending from two of the patriarch's reproductive partners were buried in the south chamber, while most
of the descendants of the other two were buried in the north chamber, along
with the few skeletal remains found of the patriarch himself—these being
in the innermost chamber. It seems, then, that the existence of the two
chambers might have been a response to the fact the clan was split between
two maternal lines, with each representing different first-generation mothers. This clearly suggests that females could play a prestigious social role in
the perceived identity of these Neolithic communities.

Aside from the patrilineal links, one male appears to be the offspring of
a clanswoman from another male outside the genetic lineage. Although
his father is missing from the cairn, this individual was buried in the south
chamber alongside a stepbrother who was a second-generation descendant
of the patriarch. This could represent an example of adoptive kinship—or
again, cuckolding—and suggests that genetic links were not the only pertinent consideration for the clan. Social—as opposed to simply biological—
fatherhood has been described in some otherwise patrilineal traditional
groups such as the Nuer.[19] The presence of eight individuals who bore no
obvious genetic ties to the rest of the clan (three of whom were women)
again suggests that factors other than biological ones could play a role in
the composition of these communities. Of course, it is much easier for us
to detect biological than social links based on power, opportunity, cultural
affinities, or other aspects. Observation of external individuals is not unique
to Hazleton; it has been observed at other megalithic sites in Ireland, the
British Isles, and France.

A plethora of radiocarbon dates suggested that the monument was built over the course of a decade between 3695 and 3650 BCE, and that its use as a funerary place probably ended around 3620 BCE. Considering the time span involved, it is likely that the cairn was in use for only few generations. It seems the wall of the north chamber collapsed at some point, and that section of the monument fell into disuse first. This event led to the disruption of burial practices in this chamber, which seems to lack additional descendants from the maternal branches. This might explain why the only fifth-generation member of the clan was found in the south chamber at a time when the north chamber was likely already blocked. This seems to be the last person from the clan to be buried at Hazleton. Afterward, the cairn appears to have been abandoned—but the remains of this remarkable Neolithic clan stayed intact until the present.

Does this mean all Neolithic societies were polygamous? Probably not. We now have ancient DNA data from a large pedigree, revealing a complex, yet slightly different, social structure. Researchers at Gurgy "Les Noisats" in France retrieved and analyzed genome-wide data from ninety-four individuals excavated from a 4700–4300 BCE Middle Neolithic site.[20] Although no associated monumental megalithic structures appeared at this burial site, the geneticists ultimately managed to reconstruct two large and contemporaneous pedigrees of a clan up to seven generations. (The first pedigree connected sixty-four individuals and the second twelve.) Eleven more individuals showed no genetic links to either of the pedigrees. As in the case of Hazleton, the results indicated a patrilocal and patrilineal society, with female mobility between groups. Female outsiders tended to be buried close to their reproductive partners, which suggests that having descendants was a primary mechanism for social integration. In some cases, it was possible to detect distant kinship affinities between women. It appears that some female descendants might have subsequently returned to the original group their female ancestors had departed from generations earlier. It is apparent that the generations were linked through the male line because fifty-one out of fifty-seven males shared the same Y chromosome lineage. Interestingly, the bones of the male founder of the first pedigree were not seen in his primary burial position, but piled up within the tomb of a woman from which no DNA data could be obtained unfortunately. Even so, it suggests she was someone important within the community.

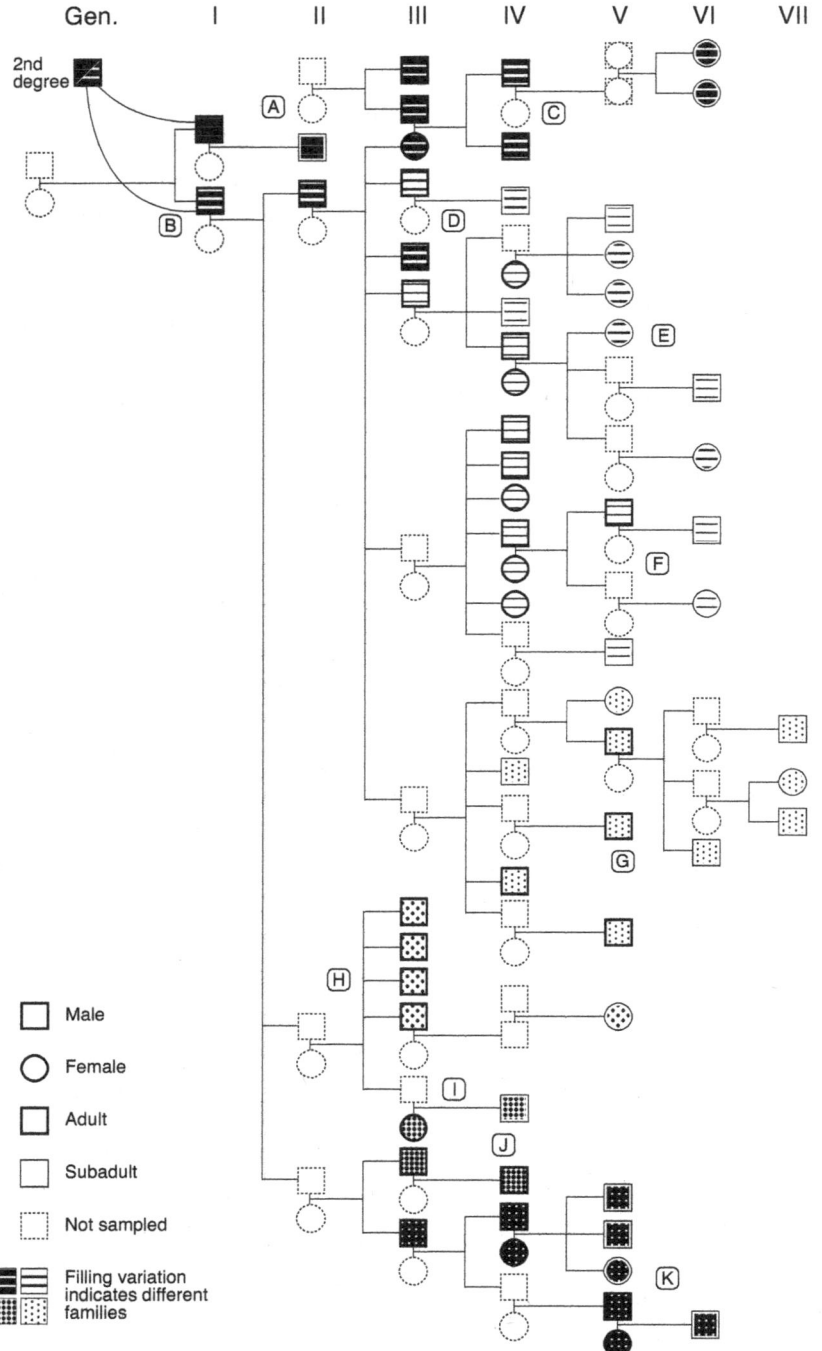

Figure 7.2

Reconstructed family tree at the Neolithic site of Gurgy (France), colored according to different families. Modified from M. Rivollat et al., "Extensive Pedigrees Reveal the Social Organization of a Neolithic Community," *Nature* 620 (2023): 602.

Although the researchers were able to reconstruct seven generations of this clan in one pedigree, as at Hazleton, the site seems to have had a rather limited historical span of just a few decades. From the observed genetic diversity, the authors estimated the clan counted some eighteen hundred members over a period of between 84 and 112 years (the higher the diversity, the larger the community would have been). In the first generations of both pedigrees there is an obvious lack of infant bones, while later on they increased. This could be evidence of worsening living conditions, perhaps associated with depletion of cultivatable terrain in the vicinity. It could be that after a few generations, the group moved on and settled somewhere else.

Unlike at Hazleton, no half-sibs were located within the group. If polygamy (or just secondary marriages) were common, we would expect to find half-brothers. Of course, it is difficult to generalize social patterns from just two examples, but it is obvious we will have more and more examples from different regions and periods in the coming years so as to fully grasp the genetic meaning of these family groups from the past.

A clan, as a kind of extended family, might seem too narrow a group to identify with in this day and age. Now people forge their collective identity—and, as a consequence, their individual profile—among larger communities that might be called tribes, populations, or nations. What can genetics tell us about these larger, suprafamily entities?

Beyond Clans

The Barbarian groups that invaded the Roman Empire after its fall are likely the social units most similar to the tribes that existed in Europe in historical times, and they served as models for the colonial powers to define human groups on other continents. There are some genetic studies on this period, including the one in the Balkans I mentioned in the preface. At the Roman military town of Viminacium, in modern Serbia, some skulls show signs of cranial deformation by a bandaging method, a non-Roman custom associated with steppe societies. We've detected the arrival of people with central/northern European and Pontic Steppe affinities—usually in admixed form—during the third and fourth century CE. The paternal chromosomes from these individuals point to these regions, which suggests a patrilineal social organization for early Germanic groups. The late antiquity Roman Empire encountered—and somehow integrated—transfrontier groups described in the contemporaneous chronicles as "Gepids" or "Goths," as supported by our results. Yet even in cemeteries conventionally identified as "Barbarian," some individuals only display the previous "local" Balkan ancestry.

In another study, this one focusing on the Barbarian migrations into the Iberian Peninsula, we sequenced more than two hundred genomes from late antiquity, sometimes in archaeological contexts associated with the incoming groups. The last day of the year 406, with the Rhine frozen, different Germanic groups that included the Vandals, Alans, and Suebi, crossed the river into Gaul and subsequently Iberia. In the Roman chronicles, it is unclear what distinguished one group from another; additional groups included the Quadi, Rugians, Heruli, Marcomanni, Lombards, Alemanni, Thuringians, and Burgundians, among others. Later on, the Visigoths—a Roman-allied Gothic tribe that undertook a seemingly incredible migration from the Balkans—entered Iberia as well, after sacking Rome under Alaric in 410. The Suebi initially settled in the northwestern corner of Iberia; under the dynamic king Rechila, they expanded south, and in 439 conquered the city of Mérida—the capital of the Roman province of Lusitania—and made it their provisional capital. Rechila's son, Rechiar, succeeded him in 448; unlike his father, he was Christian, but equally bellicose. He was the first Barbarian king to mint his own coins, with the legend *ivssv rechiari reges* (by order of King Rechiar). After being defeated by Theodoric, however, he was captured and executed, and subsequently the Suebian kingdom collapsed.

None of these developments is well documented from an archaeological or historical point of view. In recent years, though, a Suebian necropolis was discovered and excavated at Mérida, thereby offering the opportunity to analyze the ancestry composition of the people buried there around the turbulent mid-fifth century CE. With genetic data from twenty-seven people, we uncovered genetic affinities with contemporaneous inhabitants of central and northern Europe that were different from the genetic composition of the local Hispano Romans. As in the case of Viminacium, few individuals turned out to have local ancestry, including a girl buried with rich gold ornaments of Germanic style, and for this reason, she is popularly known as the "Suebian princess." She is buried in the same way and alongside people of foreign ancestry, and we can presume they customarily identified themselves as Suebi—if only culturally.

These and other pending results from elsewhere suggest that, at least during the Barbarian migrations, the concept of the tribe was a fluid one. Although it had to some extent a core of shared genetic ancestry and a common cultural background, it was able to incorporate outsiders who shared its military objectives. We may suspect that being a Suebi or even a Hun was more a job than a specific biological heritage. And yet they existed as cultural entities and were perceived as such by both their own members and Roman observers.

What about a yet higher layer of collective identity? What about the sense of belonging to a larger grouping of people, beyond the clan or tribe? Is it possible to genetically divide our species or large-scale areas such as continental regions in an objective way?

Pros and Cons of Human Population Genetics

To start exploring the pros and cons of human population genetics for understanding human identities, we might ask ourselves how diverse our species is in the natural world. If it were highly diverse, we would expect some degree of geographic structure within biological categories below the level of species (regardless of what we call them). Yet it turns out that humans are a remarkably low-diversity species as compared to others, including primates. Despite our enormous population figures, humans are identical in about 99.9 percent of their genomes.[21] This means that two randomly chosen humans differ in a tiny fraction of the nucleotides, the chemical building blocks of our genome—approximately one out of a thousand sites. This doesn't seem like much diversity, but besides the extremes (all individuals identical or all different in all genetic sites), it is difficult to grasp the real meaning of these values in a biological context. One way to look at human diversity is to compare human heterozygosity (defined as the average of gene positions that show variation within a species or population) across the genome to that estimated in other primates. The observation results in humans being close to the lowest values of all primates. (Certain hominoids such as gorillas, which suffer from dramatically declining numbers, show even lower values.)

Could it be that, though not diverse, we are a highly subdivided species? It turns out that not only are variable genetic sites relatively scarce in our species but, more important, their distribution is not strongly geo-graphically clustered. As Richard Lewontin already observed in 1972, with the limited molecular data available at that time, many of the "common genetic variants"—those found in more than 5 percent of the people's chromosomes—are shared among all human groups.[22] This also means we won't be able to define any human group because specific genetic variants are not present in any other human group.

Lewontin's observation was challenged by A. W. F. Edwards in what is sometimes known as "Lewontin's Fallacy." Edwards's 2003 paper pointed out that with sufficiently large numbers of genetic variants combined, population genetics analyses would be able to classify any given individual correctly to their population of origin.[23] A type of multivariate analysis of genomic data called principal component analysis provides a fairly accu-rate representation of the geographic map of modern Europe, meaning genetic information is to a large extent geographically structured, even at a regional scale. Some researchers suggest that these results do not invalidate Lewontin's original observation per se, nor do they support the existence of race-style subdivisions. Most genetic variation is still distributed among individuals within populations. In fact, I think both researchers are arguing about different observations, not mutually incompatible. The fraction of genetic variation explained in these representations is always low, in the range of less than 10 percent with the first two principal components com-bined. There is still enough information, however, to be able to attribute individuals to their populations or countries of origin with a high level of accuracy.

Another challenge has to do with visualizing the magnitude and extent of human genetic variation. Traditional representations like genetic trees shed no light on this crucial aspect of human genetics. Recently James Kitchens and Graham Coop proposed to represent it as areas of circles.[24] Starting with about 2.9 billion measurable nucleotide sites that can be vari-able across the human genome, they looked at a sample of 609 individuals from the Americas and found a figure of only 39 million real observable variants. If they restricted the analysis to those genetic variants defined as "common," the figure decreased to just 10 million sites, an almost insig-nificant number compared to the potential general one. The researchers repeated the approach for different populations, invariably finding remark-ably low numbers of variable sites as compared to the total length of the genome. For instance, 99 Utah residents with northern and western European ancestry showed 5,726,377 variants, and 96 African Caribbeans from Barbados showed 8,018,649 (the latter likely reflects the larger genetic diversity found in Africa). More important, between populations the level of overlap in common variants is enormous, which again means human groups have few exclusive genetic variants. In fact, variants that are common in a single geographic region are usually not exclusive to that region but instead present at certain levels in other world populations.

An additional interesting observation is that if two individuals from the same region differ genetically, it is often due to differences in genetic variants that are found globally.

We've seen that the notion of races has been largely discredited. But what about similar classifications we routinely use nowadays in population genetics papers? Are they really any more meaningful than races? Or do they follow a similar pattern—that is, being a construct and self-supporting their existence by common usage? The innovative works of Italian geneticist Luca Cavalli-Sforza showed human genetic variation as bifurcating phylogenetic trees, representing similarities via arbitrarily chosen population labels such as "Caucasoid," "North Eurasian," and "Northeast Asian." These branching diagrams usually depict lines of evolutionary descent from a common ancestor. Working with limited serological data, the only molecular information available prior to the advent of genomics, Cavalli-Sforza tried to deduce from his trees which of the major subdivisions in our species—without calling them races—clustered together. He concluded that Europeans and Africans had that distinction, although we now know that all modern non-African human groups are more similar to each other than any of them is to sub-Saharan Africans.[25] Moreover, Cavalli-Sforza's trees could not legitimately represent the history that produced the observed pattern of genetic similarity.[26] Phylogenetic trees generated with intraspecific genetic data are misleading representations because branches that separate before others might suggest the existence of "more and less evolved" populations within a single species, which is a wrong inference of the evolutionary time.

Phylogenetic trees are also sensitive to the choice of populations and the samples that are supposed to represent these populations. In some ways, the choice for human populations can be compared to the old races. (Depending on the samples incorporated, one can find what one expects to find

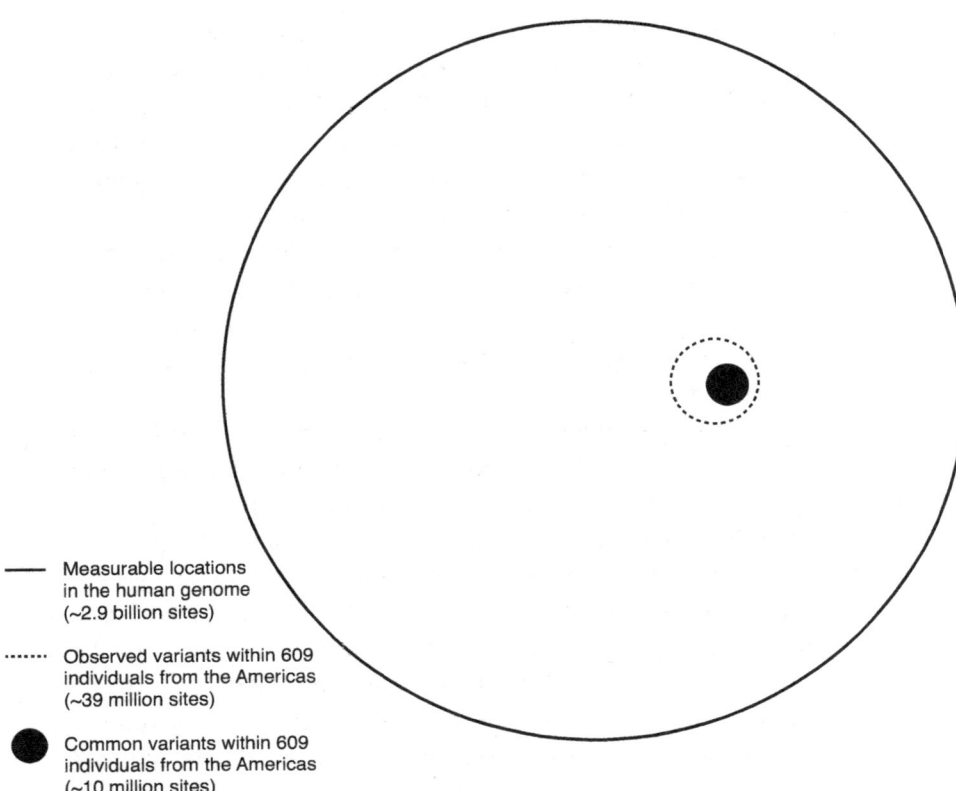

— Measurable locations
in the human genome
(~2.9 billion sites)

······ Observed variants within 609
individuals from the Americas
(~39 million sites)

⬤ Common variants within 609
individuals from the Americas
(~10 million sites)

Figure 7.3
Circles showing the relative magnitudes of potential variable genetic sites in our
genome (solid circle) versus observed genetic variants in a sample of 609 individuals
from the Americas (dashed circle) and common genetic variants (found in more than
5 percent of the people's chromosomes) in the same sample. Modified from J. Kitchens and G. Coop, "Visualizing Human Genetic Diversity," *James Kitchens* (blog), May
16, 2023, https://james-kitchens.com/blog/visualizing-human-genetic-diversity.

because the selected categories create a sense of structure that might not
actually exist.) Races were never a formal scientific concept, but some population labels suffer from the same problem and have been progressively
abandoned as well. Referring back to the archives of the *American Journal of
Human Genetics*, the use of "Caucasian" in papers has declined from 12 percent around the mid-twentieth century to less than 1 percent at the beginning of the twenty-first century. Alongside this decline, continental labels

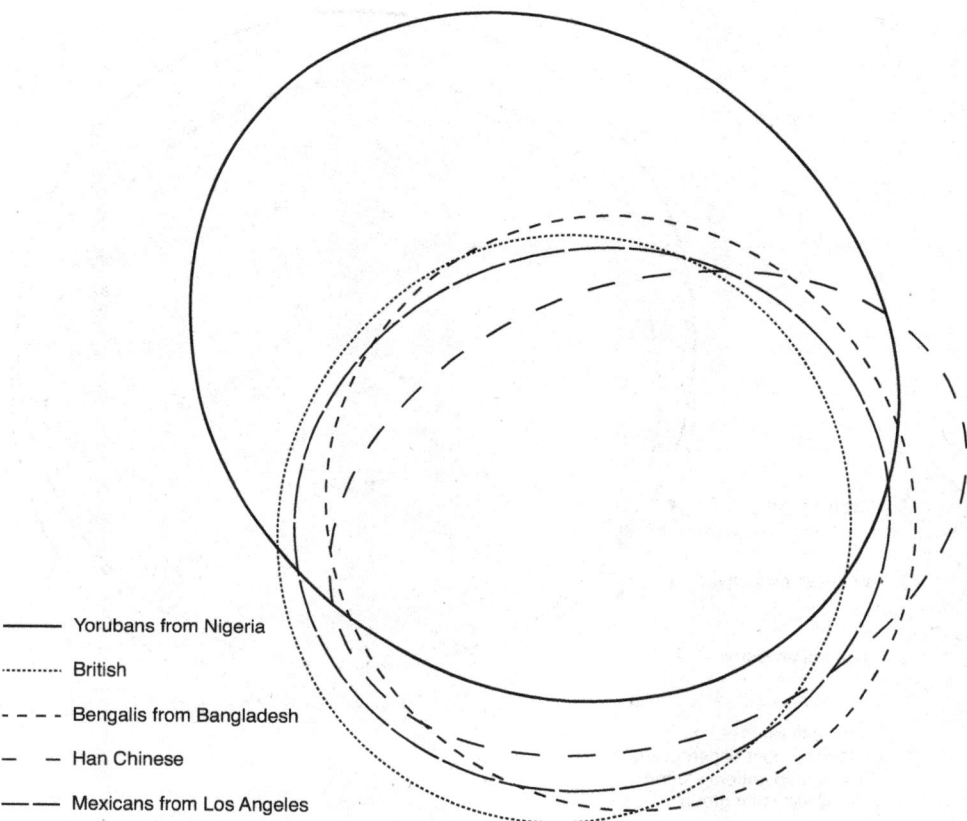

Yorubans from Nigeria

British

Bengalis from Bangladesh

Han Chinese

Mexicans from Los Angeles

Figure 7.4
Overlap in common variants between human groups: Yorubans from Nigeria, British, Bengali from Bangladesh, Han Chinese, and Mexicans from Los Angeles. The most diverse group is the one with African ancestry. Modified from J. Kitchens and G. Coop, "Visualizing Human Genetic Diversity," *James Kitchens* (blog), May 16, 2023, https://james-kitchens.com/blog/visualizing-human-genetic-diversity.

have increased—for instance, the use of "European" surged to 42 percent of papers between 2009 and 2018.[27]

What is a human population then? To start with, the definition is not at all obvious or easy; it can in fact be expanded or contracted at will and is notoriously difficult to pin down in an objective way.[28] In population genetics, a population is operationally defined as a group of individuals that mate with each other and are to some extent different from other groups.

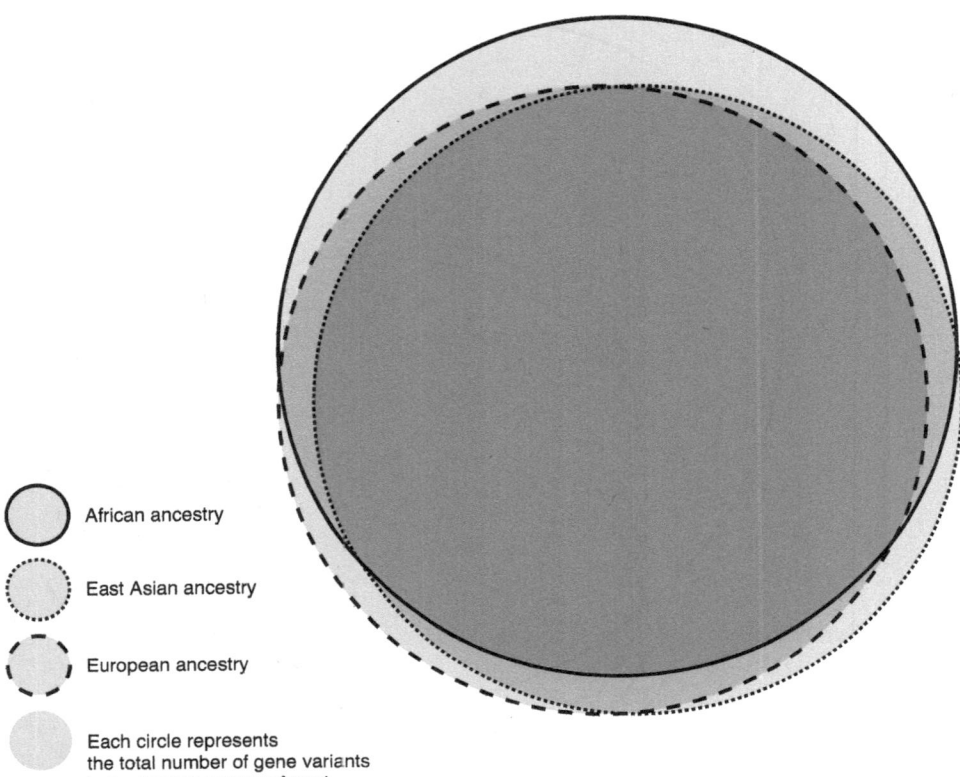

African ancestry

East Asian ancestry

European ancestry

Each circle represents
the total number of gene variants
in the human genome found
within a particular group

Figure 7.5
Genetic differences within and between groups. Groups are genetically alike because
they share the same sets of within-group gene differences. A human subdivision in
large groups (namely races) would only be possible if these circles did not extensively
overlap. Modified from B. M. Donovan et al., "Toward a More Human Genetics Education," *Science Education* 103 (2019): 533; adapted from N. A. Rosenberg, "A Population-
Genetic Perspective on the Similarities and Differences among Worldwide Human
Populations," *Human Biology* 83, no. 6 (2011): 665.

These populations are sometimes defined by their cultural or geographic traits (for instance, speaking a different language than neighboring groups, or being restricted to an island or to the boundaries of a nation). In the case of humans, there are no truly isolated populations and this poses problems for our definition for a number of reasons, including the porous nature of the political boundaries of modern states. In some analyses, scientists have chosen to work solely with nationalities, and in others, cultural attributes. But there are no good solutions. (Imagine trying to compare individuals from small states like Luxembourg with geographically immense and diverse Russia, for example.) Indeed, the different definitions of populations depend on the questions we ask for a particular genetic dataset.

In the natural world, well-defined genetic populations can be found. There are certain geographically restricted populations or species that are isolated from others—think, say, of cases of insular endemism like the extinct dodo from Mauritius island. In these cases, the largest population is the limit of a species. But such examples are impossible to find in modern humans; as we've seen in previous chapters, we are all genealogically connected to some extent as we go back in time. There have been adaptations to specific environmental conditions in the last tens of thousands of years, which means a certain degree of geographic structure of genetic variation can exist, typically in pigmentation or pathogen-resistance genes. Still, such patterning is deceptive because few genes in the total genome are associated with phenotypic features. One explanation for the perception of race is that pigmentation genes are a kind of exception in human diversity due to a certain geographic patterning. The apparent geographic structure erroneously put forward by racial scientists has been disproven by genomics with the analysis of whole genomes worldwide, which show not only genetic cline patterns across vast geographic areas (after all, races are defined as categorical, not gradational entities) but also genetic gradients that are different from those observed in pigmentation genes.[29]

Moreover, it can be statistically demonstrated that clusters of populations resembling past races can be generated from DNA data simply by sampling discontinuously.[30] For example, if we were to generate a subjective population such as "Atlantic Coast Europeans" (joining Portuguese, Spanish, and French individuals from the Atlantic coast) and compare it to another subjective population such as "Adriatic Coast Europeans" (joining Italians, Croatians, Montenegrins, Albanians, and Greeks from the Adriatic coast), we would surely find genetic differences between the two groups—which is not at all to suggest they are "natural" populations. An additional problem is the way in which ambiguous and contradictory labels such as "Hispanic," "Mexican American," "European," and "white" are perceived by the general public.[31] And yet many population genetics researchers are using similar concepts. Of course, populations are necessary for most of the analysis performed in population genetics, not only in humans but in the rest of the species of the natural world (where there are consequences in conservation policies, for instance). This doesn't mean that a population has to be an obvious natural phenomenon, such as a totally endogamous group with not a single individual mating outside the defined community.

It isn't just population geneticists who are using ambiguous population labels; ancestry companies, too, use geographic categories to present their results to customers in ways that can be misleading or confusing. One can be categorized as "southern European," "British Irish," "Greek and Balkan," or just "sub-Saharan African" (the latter is an enormously diverse category that likely reflects the poor sampling of customers on the African continent). But just as in population genetics studies, the population labels selected a priori to some extent determine the perceived outcome of the analysis, as we saw in chapter 1 with my "France and Germany" personal heritage. In addition, the accuracy of the results would be strongly affected by the chosen reference population datasets.[32] As well, a geographic connection to present-day individuals in a particular country can be misleading because the genetic similarity could in fact indicate a shared ancestry hundreds of years ago in a different region. Ancient genomic results clearly suggest that migrations have been prevalent; with the use of these population categories, however, we tend to underscore the last few hundred years of human mobility and create a false impression of genetic fixation.[33]

Not only are population labels and phylogenetic trees problematic; any visual representation of genetic diversity is likely to mislead to a certain degree. Even the usual representations employed in population genetics papers, such as principal component analysis, graph models, and admixture analysis, are not free of bias. Still, data need to be visualized to be interpreted. Probably the best option would be to represent each dataset with a different visualization tool in order to get a broader frame of interpretation.

Ancient genomics can provide direct information on the genetic composition of some of the ancestors of modern populations. Nevertheless, the new subfield of population genetics also uses terms that are imprecise or potentially misleading, such as "early farmers' genetic ancestry" or "East Asian genetic ancestry." Such terms are subjective choices made by researchers and they are often interpreted differently by other researchers—not to mention the general public. As these labels are based on statistical methods, they may seem more objective than previous ways to study genetic diversity based on cultural traits or geographic features. But genetically described populations are also constructs—in this case, modeling constructs. Just as in modern population genetics, they are highly dependent on the reference populations used in the statistical analyses. And in the case of ancient genomics, they are often based on ancient genomes from specific archaeological horizons and chronological periods that are supposed to represent a particular ancient population (which might be labeled "Steppe nomads," "Romans," "Slavs," or "Anglo-Saxons").

Ancestry can be defined historically, culturally, geographically, or religiously as well as in purely genetic terms, and some of these definitions can be contradictory.[34] Genetic ancestry is a concept that captures people's imagination and for this reason has been enthusiastically used by ancestry companies. Though their customers may be thrilled to know they carry "Viking" or "Yamnaya" ancestry, such terms are meaningless in historical, cultural, or even genetic terms. Ancestry itself is rarely defined, and individual results can thus be overinterpreted.[35] Some may consider it relevant to their identity to have an ancestor within a particular category, but they are likely to misinterpret the information. To have ancestry found in Viking genomes does not makes you a Viking.

Genetic ancestry is in fact a quantification of genetic similarity to reference populations. Therefore according to geneticists Iain Mathieson and Aylwyn Scally, "Most statements about ancestry are really statements about genetic similarity, which has a complex relationship with ancestry, and can only be related to it by making assumptions about human demography whose validity is uncertain and difficult to test."[36] Coop likewise argues that it is more accurate to refer to "genetic similarity" than "genetic ancestry" in describing samples and that population genetics should move in this direction, especially now that genomic sampling of humans is increasing.[37]

Coop also thinks "similar to" doesn't imply "identical to"; it remains to be seen whether this subtle difference is acknowledged by the public. The

same issues apply to the ancestry companies, which are compelled to provide engaging and easy-to-digest stories to their customers, often at the cost
of misinforming them about the science behind the stories.[38] That said, if
an analysis is properly done, shared ancestry between a customer and, say,
the Birna Viking warrior would in fact mean that both individuals descend
from a population that at least partially had a similar genetic component.
But one additional observation should be kept in mind: to some extent,
we are all descendants of a Viking, or Charlemagne, or numerous others
because we are all connected—and recently so.

One positive result that has emerged from ancient genetics studies is that
human groups have no single ancestry component but rather several layers
of ancestry, one imposed on top of the other across time, as a consequence
of different migration events. In fact, we should speak of "ancestries," in
the plural, to describe any individual or population.

The complex interplay between different layers of ancestry and population weakens the general concept of races as real biological entities.[39]
Instead of having a common and relatively uniform genetic background,
shared with other people from the community we think belong to, what
actually happens is that different populations—however we define them—
have slightly different proportions of the different ancestry components.
Rather than constructing abstract past populations, ancient genetics offers
us the possibility to refer to specific sites and archaeological horizons where
a certain set of skeletons originated. It is something much more subtle and
thus more difficult to encapsulate as a comprehensible sign of group identity. I think this is a more positive—albeit complex—view for building collective identity levels from the past within our species.

The Limits of Our Species

Actually, now that we've retrieved genomes from ancient hominins such
as Neanderthals, we're not even sure about the biological limits of our
species. The biological definition of a species (members of a species can
interbreed and produce fertile offspring) remains subjective, among other
things because until recently we weren't able to analyze ancient genomes
from closely related evolutionary lineages. When we did so, we discovered
that modern humans, Neanderthals, and Denisovans (the latter two being

species or populations inhabiting western and eastern Eurasia, respectively, during the Middle Pleistocene) all interbred at different moments and in different places. This has created a new paradigm for recent human evolution. Instead of a branching tree, what we have is a complex pattern of interconnections, like a capillary system, that it is much more difficult to separate into different entities—whether they're called populations, lineages, or species.

Furthermore, it seems quite likely all three lineages would not have been able to see each other as essentially different (an additional difference is that Denisovans have been discovered genetically while Neanderthals have a diagnostic anatomy known for more than 150 years). We know from genetics that they were organized in patrilocal, small family groups, and I'd bet this was their basic identity horizon. Saying that Neanderthals existed and identified themselves as a collective—and one different from others such as modern humans—is again a process of reification. Probably our interpretation of prehistory tells us more about ourselves and our ideological agenda than about ancient humans, who are after all extinct.[40] Perhaps our biological definition of species has simply not yet embraced the notion of parsing genetic data in a temporal dimension.

Therefore the genetic limits of our collective identity may even go beyond all humankind. It could have been larger in the past, with no objective boundaries at any given moment. Our evolutionary history is best imagined as an entangled network of interconnected ancestors shaping the stream of heredity. We may break down this continuous evolutionary flow into segments to name different hominin species (*Homo habilis*, *Homo ergaster*, *Homo erectus*, etc.), sometimes represented by just a few cranial remains. This process might be partly influenced by the paucity of the fossil record. Here again, though, such paleontological species are subjective and sometimes follow an ideological narrative. This should come as no surprise; as in the case of population, there are many definitions of species and their use depends on whatever questions we want to ask.

Criticizing the grouping of humans is not the same as saying that human variation does not exist. It does exist, and to some degree is externally visible and geographically diverse. It's just that it isn't structured as we superficially perceive it. Although the message is easy enough to understand, the underlying difficulties historically experienced due to the use of category levels in population genetics need to be given thoughtful consideration. Despite the operational need to use populations, no clear alternatives present themselves.

And despite the problems and limitations I've described, genetics is having an enormous impact on the understanding of the complex historical processes associated with migrations and changes in ancestry that have shaped our genomes, definitively linking the past to the present on all continents. Genetics is leading an intellectual revolution in our view of the temporal dimension of our collective identity and will continue to influence the future of human studies. We are living truly exciting years of scientific discovery at an unprecedented scale, yet we need to understand what these studies are telling us in terms of collective identities.

8 The Future of Human Identity

A man is the sum of his actions, of what he has *done*, of what he can do. Nothing else.

—André Malraux, *Man's Fate*

The 1982 cult film *Blade Runner* shows a dystopian future where genetically engineered humanoids designed to live for four years are used as slave labor in extraplanetary colonies. Although they are no different physically from real humans, they are not considered legally human and can be hunted down by special police units called "Blade Runners" if they try to escape. The film poses a range of interesting ethical and philosophical questions, one of which refers to the replicant's search for identity. Being created with no past and therefore no infancy, some resort to keeping old family portraits to help give them false personal memories from childhood. *Blade Runner* illustrates an obvious fact: our identity is to a large extent rooted in our past memories and own vital trajectory. The replicants want to have a past as much as they want to have a future. And they would like their memories to be remembered and not "lost in time, like tears in the rain." Similarly, the central drama of dementia is experiencing memory loss that fades one's notions of identity.

With the arrival of new technological developments, we might ask ourselves, How will human identity evolve in the future? We've been exploring human uniqueness throughout this book. But what about human-made copies? What would it mean to be genetically identical without sharing memories? And what would that imply in terms of identity?

Human Clones

On February 23, 1997, the news leaked that the journal *Nature* was about to announce the first mammal clones from adult cells of another individual—in this case, a sheep who was going to be named Dolly. I was living in Cambridge at the time, and I remember that day. The world was both shocked and fascinated, and everything seemed technologically possible. We felt the future was knocking at our door. Although the clone was a sheep, everyone grasped there was nothing to prevent the same thing being done with humans. Governments and ethical experts alike rushed to ban human cloning, even though some collateral aspects of the technique could be useful for assisted reproduction.[1] And, rather surprisingly, twenty-five years later, not only have humans not been cloned but, apparently, nobody has even attempted it. In the meantime, technical difficulties and the overall inefficiency of the process as well as potential abnormalities in animal clones have emerged as drawbacks. Still, the desire to generate cloned humans has not entirely subsided.[2]

I think the issue with human cloning can be linked to our perception of human identity as being reduced to our DNA—a misperception, as we've seen, based again on notions of genetic essentialism. In fact, if we were to produce a clone from an adult person, the resulting copy would be physically similar but most likely different in all other respects since the environment—including diet, health, education, and so on—would have changed enormously. And as in *Blade Runner*, the clones' own, new memories would make them entirely different people. I am almost sure that the clone of a singer, a painter, or even a scientist would not necessarily turn out to be a singer, painter, or scientist. The process would be like formatting a computer's hard drive and restarting. If anything, it could be argued that cloning a specific person with a specific talent (think, for instance, in a basketball player) could prevent the clone from exploiting possible, alternative talents.

An example of how the outcome of cloning may be misunderstood is provided by pet cloning, which has recently emerged as a successful business. Many pet owners want to have a perfect copy of their deceased, beloved animal. The first cloned dog was produced in 2005 by a team of South Korean researchers; they cloned an Afghan hound called Tai from the skin of a dog's ear and produced two identical puppies. Although one died soon after birth, the second, named Snuppy, lived a healthy ten years.[3] There are now several companies that clone dogs and cats, charging around fifty thousand and twenty-five thousand dollars, respectively. (Most cats, I suppose, would be dismayed to know they're cheaper.) Actress and singer Barbra Streisand famously cloned her dog Samantha and was given back five puppies, from which she kept two "neo-Samanthas." As she explained in a *New York Times* piece, "Each puppy is unique and has her own personality." She concluded that you cannot clone the soul.[4] Not surprisingly, the personality of a dog—or person for that matter—is not just their DNA but also the result of the interaction of their genome with an ever-changing environment—and, I would add, with a certain degree of randomness that shapes our lives. If anything, this story again demonstrates the misconceptions about the power of genetics to form identities.

Editing Humans

If individuality is represented by our unique genome, what would happen to our identity if we are able to modify our genes? Another scientific development may shed new light on the future of identity: the precision tools provided by CRISPR-cas9 to modify specific positions of our genome. CRISPR-cas9 has only been applied once to human embryos—rather controversially, as the person who did it, Chinese expert He Jiankui, ended up in prison for three years.[5] Still, it seems it may soon become socially acceptable to use it on the human germ line, at least in life-threatening situations. Despite the problems associated with unintended editing outcomes in as many as 16 percent of embryos, in the future it could become a convenient tool for changing certain traits in our genome, especially those that depend on few genetic variants.[6] For instance, people might want to improve disease resistance in their offspring or even modify physical aspects that could make them more socially attractive. In particular, fair eye color depends on only a half-dozen genetic variants, but at least one of those, located in a regulatory element close to the HERC gene, is crucial for blue eyes.[7] Interestingly, ancient genomic analysis has demonstrated that most European foragers prior to the arrival of the first farmers from the Near East about eight thousand years ago had a unique combination of blue eyes and dark skin pigmentation.[8] This phenotype is long gone and remains as an ancestral layer that only reaches 35–40 percent of modern European genomes—and this number is only found in Scandinavia. The underlying genetic variants are still among us, however, and could be restored by gene editing.

Although it remains purely speculative, the tools now exist to artificially modify our genetic individuality—and consequently, our sense of identity. Still, as we've seen, it would be difficult to increase our sense of uniqueness by artificial means since each of us is already unique.

Human Genetics and Traditional Views of Identity

Our genetic identity would make sense as a symbol of a unique individual makeup. At the collective level, it would make sense as a unique member of our species, with a certain degree of geographic attribution. Nevertheless, this concept can apply to members of any other species; it is not exclusive to humans. In fact, almost any other species in the natural world—including other primates—seems to harbor higher genetic diversity than ours. Therefore while genetically unique, humans are limited in "uniqueness" compared to other animals.

When looking at the individual human genome, we face a paradox: if identity is rooted in genetic diversity, then there seems to be too much genetic variation at the individual level, making it difficult to interpret in terms of personal values and aptitudes. Genetic individuality is not equal to personality; if anything, it is a paragon of uniqueness. While it is true that genetics supports our individuality, as it encodes such elements as our skin color, height, and disease resistance or susceptibility, all of these traits have a complex role in forming our sense of collective identity too. In many ways, genetics confirms the famous quote by Michel de Montaigne is his *Essays*: "Every man carries the entire form of human condition."[9]

Moreover, people cannot easily use genetics to identify themselves as members of any objective category with sharp boundaries. The reason for this stems from the complex definition of "population" in population genetics (note that traits determined by social, political, and cultural factors can be equally complex). Despite a potentially large range of genetically variable sites across the genome, human diversity largely overlaps among populations while being poorly geographically structured between them. Human groups in which all the members share specific genetic variants *not present* in other groups simply do not exist. By retrieving ancient genomes, we can delve deeply into the past to see how major migrations and recurrent admixing form the basis for the current diversity. As we go back in time on all continents, layers on layers of genetic ancestry are being revealed. It is difficult to build a sense of collective identity by looking at genetic components we share with someone from the distant past, even if such analysis proves to be scientifically sound.

Genealogical genetics can be equally disappointing from the perspective of psychological essentialism. Not only is the likelihood we've inherited a significant fraction of one of our ancestors low after just a handful of generations, but it soon becomes evident that we are all distantly related. To make things even knottier, our genetic ancestors form just a small fraction of our genealogical ancestors. Who is more important to our sense of identity, a genealogical ancestor or a genetic one? And without the genomes of everyone involved, how can we know who is who?

As we've seen, genetic similarity, genealogy, and genetic ancestry are closely related concepts, and yet their complexity defies easy comprehension.[10] Genetics cannot provide a simple and comprehensive way for us to define ourselves as members of a collective with clear delimitations. Even identical twins are somehow slightly different from a number of perspectives, including that of genetics itself—perhaps paradoxically, being "identical." And unless we are related, facial resemblances do not reflect genetic similarity either, again leaving us without any objective way of linking genetics to identity. The discovery that some people are mosaics of female and male cells is just one example of how genetics, rather than resolving any current debate in simplistic fashion, only reveals further complexity.

Toward a New Concept of Genetic Identity

The irresolvable nature of identity poses a problem for humankind. Genetics holds great promise for consolidating human views on identity less controversially than traditional arguments, which mainly revolve around cultural traits and physical aspects. This now seems overly optimistic as new complexities emerge from genetics, both in present-day and ancient humans. Some studies go so far as to suggest that genetics lacks the power to rewrite the definitions of collective identities and instead tends to reinforce the existing definitions of race and nation, banal and politicized as they may be.[11]

As sensitive information about human nature permeates all layers of society, it is processed by the general public, including policymakers, in many different ways due to a dearth of clear messages and consensus. While many realize that genetics can explain fundamental aspects of human nature, the complexities associated with this scientific endeavor are often overlooked and misunderstood, as we've shown in this book. Not only does official science face difficulties; it's obvious that ancestry testing companies can be much better at explaining how their customers should interpret the results that can be so crucial to their personal identities.

That said, if properly explained, genetics can be a genuine transformative force in both individual and collective identities. Genetics shows that we are all interconnected, just as it reveals how every person in humankind is unique. It demonstrates that contemporaneous genomes are a composite of different and widely shared layers of ancestry, and helps us understand how the past has shaped our genomes. It provides information from historically marginalized groups that can now be restored to the human family through evidences of shared genetic links with both the past and the present. These observations are by no means trivial; in fact, such powerful ideas can have a considerable impact on society. If properly communicated, genetics can be a democratizing force in the twenty-first century.

Naturally, there are tensions between the idea of having an innate layer of identity, as demonstrated by genetics, and the postmodern construct of identity as a composite of multiple, fluid, and even contradictory selves.[12] There are many definitions of identity and undoubtedly genetics will contribute to yet others in the twenty-first century. It remains to be seen how these will mesh with the previous notions based primarily on such nongenetic factors as language, culture, education, social status, gender, and religion (some of which, interestingly, are also highly heritable). Definitions based on gender, population, racial identity, and the like should be adopted by social consensus, owing to their inherent complexity and subjectivity. By doing so, we can help make the human biological sciences a more positive social institution and leave behind its problematic associations with race theory and eugenics.

All notions of identity, whether provided by family records or ancestry companies, are based in the end on a set of subjective assumptions. How we deal with this need to know what DNA can tell us depends on our inclinations, and there are different options available, from the rise of populism to the emergence of a humankind-based identity. As we've seen, patterns of human genetic diversity fail to support the conventional categories attributed to our species. And yet collective identity bears a surprising political solidity capable of bonding thousands or even millions of people across the planet, and has become a primary category for self-definition.[13] But not being an objective truth, group genetic identity becomes a fragile concept prone to misuse and falsehood. I believe that with the information at hand, we need to be wary of those brands of populism that frame identity in exclusionary terms.[14]

Most of us will be forgotten in two or three generations. Even within the family context, we'll soon end up as no more than the subjects of anonymous photographs, or names with dates of birth and death in some future genealogy. Perhaps the geneticists of the future will uncover fragments of our genome scattered among various individuals elsewhere in the world. As long as humanity exists and has the capacity to remember, an endless stream of heredity will connect us to the past and the future. But our real impact will most likely be what we do in life instead of what we are supposed to be by birth, what we do for our family and peers instead of whichever chromosome blocks we transmit to subsequent generations. The ways we influence our descendants as well as our friends and contemporaries are probably more relevant than the bunch of genes we happen to send off to the future. Our identity, the one that makes a real difference to the world, is in our hands, not in our cells. True, it is a bigger responsibility, but it's also more democratic. (Among other things, you don't need to have children to have a positive impact on society.) Yet trying to define ourselves by our actions is harder work than relying on an external observer of group identity; for one, it demands we take action.

The decline of religion since the eighteenth century has triggered the rise of ideologies, which in turn face their own demise in a globalized and technological world, leading to a rise in individuality. I suspect that collective identities will experience further social decline in the twenty-first century. Recent polls in the United States estimated that 79 percent of Gen Z—those born between the mid- to late 1990s and early 2010s—are lonely as compared to 50 percent of baby boomers. This could reflect a negative side effect of the individuality trend, alongside the rise of secularism, a drop in marriage rates, and a failure to achieve military recruitment goals in the United States and other Western countries. (The church, family, and army are three powerful engines of collective identity in any Western country.) It is entirely possible that future collective identities will be based more on social status than supposed shared history and ancestry.

The weakening of traditional collective identities does not mean they cannot coexist with individuality. I think it means they need to be reformulated in a more inclusive and scientifically informed way. Maybe this would be a good time to promote a version of collective identity represented by humankind. A humankind that embraces all identities can help us face collective challenges, such as the climate crisis that threatens our very survival. This is not the same as advocating for a global community that transcends cultural group differences but instead simply acknowledges it as a legitimate additional option for identity. One potential difficulty is that collective identities like nations and religions invariably hold the promise of a brilliant, limitless future—a promise that may be much more limited at the individual level, especially in a world of growing inequality. Future notions of identity will need to resolve the tensions between individual and collective identities, and genetics can play an effective role in this task.

In the coming age of individuality, all humans, one by one, unique in their diverse genomic identities, connected to each other in an immense global network of collective identity, will be needed in the gigantic effort to ensure the survival of our civilization. The way forward will have to be built, and it will require a new type of identity that should link the past to the future—one in which genetics has a positive role to play. The past is in our genes, but the future is in our hands.

Acknowledgments

Getting close to the age of my late father, I look at the mirror and see him: his nose, eyes, eyebrows, and even his wrinkles. Besides having half of his genome, am I like him? Not really, because his personal circumstances that included fighting as a teenager in the Spanish Civil War were quite different from the circumstances I grew up in myself. To name but one difference: the fact of having him as a father. People are their genetic background plus all the things that happen around them, including family and friends. People are unrepeatable and irreplaceable. And my wife, Marta, and my kids, Martina and Marc, are an irreplaceable part of my identity.

I am thankful to multiple friends who gave me feedback, not only because they shaped the book itself but because they keep shaping me one way or another. We are what we are and what we try to be. We are also our work.

Agnar Helgason, who has a unique background in genetics and social sciences, was a driving force in the early version of this book. Simón Parera, a researcher and LGBTQ activist, kindly illuminated me on the complexities of sexual and gender identity. David Reich, a friend and collaborator as well as world-leading paleogeneticist, is a constant intellectual reference for my work. Rogers Brubaker, Wendy Roth, Melanie Griffiths, Oscar Vilarroya, Mike Grbic, Tomàs Marquès-Bonet, Iñigo Olalde, Vanessa Villalba-Mouco, and Arcadi Navarro provided useful remarks and comments on some aspects of the manuscript. In the last years, I have also been in contact with genealogical and genetic researchers such as Maarten Larmuseau, Francesc Calafell, and Francisco C. Ceballos.

I am indebted to Daniel Schechter and Cindy Milstein, who provided invaluable editing of the whole manuscript, and Quim Massana, who created the figures. I am also grateful to the Museum of Natural Sciences of Barcelona and the Consejo Superior de Investigaciones Científicas of Spain for institutional support along with the Spanish Ministry of Science, Innovation and Universities for funding my research on paleogenetics.

I also would like to thank Anne-Marie Bono, the MIT acquisitions editor, for her continuous support and enthusiasm on this project. Without her, this book wouldn't exist, I think.

Notes

Preface

1. C. Lalueza-Fox, *Inequality: A Genetic History* (Cambridge, MA: MIT Press, 2022).

2. L. Scott, "It's the End of Globalization as We Know It (and That's Probably Fine)," *Dispatch*, March 23, 2022.

3. J. Mankoff, "Russia's War in Ukraine: Identity, History and Conflict," Centre for Strategic and International Studies, April 22, 2022.

4. S. Mallick, A. Micco, M. Mah, H. Ringbauer, I. Lazaridis, I. Olalde, N. Patterson, et al., "The Allen Ancient DNA Resource (AADR): A Curated Compendium of Ancient Human Genomes," preprint, BioRxiv, April 6, 2023, https://www.biorxiv.org/content/10.1101/2023.04.06.535797v1.

5. I. Olalde, P. Carrión, I. Mikić, N. Rohland, S. Mallick, I. Lazaridis, M. Mah, et al., "A Genetic History of the Balkans from Roman Frontier to Slavic Migrations," *Cell* 186, no. 26 (2023): 5705–5718.e13.

6. M. Samorukov and V. Vuksanovic, "Untarnished by War: Why Russia's Soft Power Is So Resilient in Serbia," Carnegie Endowment for International Peace, January 18, 2023.

7. L. B. Jorde and M. J. Bamshad, "Genetic Ancestry Testing: What Is It and Why Is It Important?," *Journal of the American Medical Association* 323, no. 11 (2020): 1089–1090.

8. W. Roth and S. Yaylaci, "Genetic Options and Constraints: A Randomized Controlled Trial on How Genetic Ancestry Tests Affect Ethnic and Racial Identities," *American Journal of Sociology* 4 (2024): 1172–1215.

9. N. Golbeck and W. Roth, "Aboriginal Claims: DNA Ancestry Testing and Changing Concepts of Indigeneity," in *Biomapping Indigenous Peoples*, ed. S. Berthier-Foglar, S. Collingwood-Whittick, and S. Tolazzi (Leiden: Brill, 2012), 415–432; R. McGreevy, "US Academic Believes He Is the First Person to Gain Irish Citizenship Based on DNA Test," *Irish Times*, July 21, 2024.

Chapter 1

1. B. D. Tatum, "The Complexity of Identity: 'Who Am I?,'" in *Readings for Diversity and Social Justice: An Anthology on Racism, Sexism, Anti-Semitism, Heterosexism, Classism and Ableism*, ed. M. Adams, W. J. Blumenfeld, H. W. Hackman, X. Zuniga, and M. L. Peters (New York: Routledge, 2000), 9–14.

2. C. Taylor, "The Politics of Recognition," in *Multiculturalism: Examining the Politics of Recognition*, ed. A. Gutmann (Princeton, NJ: Princeton University Press, 1994), 25–73.

3. M. Griffiths, "Identity," *Oxford Bibliographies*, June 29, 2015, https://www.oxford bibliographies.com/display/document/obo-9780199766567/obo-9780199766567 -0128.xml.

4. W. E. Thomas, R. Brown, M. J. Easterbrook, V. L. Vignoles, C. Manzi, C. D'Angelo, and J. J. Holt, "Social Identification in Sports Teams: The Role of Personal, Social and Collective Identity Motives," *Personality and Social Psychology Bulletin* 43, no. 4 (2017): 508–523.

5. B. Anderson, *Imagined Communities* (London: Verso, 1983), 51.

6. R. Brubaker and F. Cooper, "Beyond 'Identity,'" *Theory and Society* 29, no. 1 (2000): 1–47.

7. A. Nordgren, "Genetics and Identity," *Community Genetics* 11, no. 5 (2008): 252–266.

8. A. Nelson, *The Social Life of DNA: Race, Reparations, and Reconciliation after the Genome* (Boston: Beacon Press, 2016).

9. R. A. Bentley, "Prehistory of Kinship," *Annual Review of Anthropology* 51 (2022): 137–154.

10. J. Diamond, *The World until Yesterday: What Can We Learn from Traditional Societies?* (London: Penguin Books, 2013).

11. Brubaker and Cooper, "Beyond 'Identity.'"

12. S. Pinker, *The Blank Slate: The Modern Denial of Human Nature* (London: Penguin Books, 2002).

13. K. Marx, "Theses on Feuerbach," in *Karl Marx and Frederick Engels: Selected Works*, vol. 1 (Moscow: Progress Publishers, 1969), 15.

14. J. P. Sartre, *Existentialism Is a Humanism* (London: Methuen, 1957), 28.

15. D. P. McAdams, "Personality, Modernity, and the Storied Self: A Contemporary Framework for Studying Persons," *Psychological Inquiry* 7 (1996): 295–321.

16. Pinker, *The Blank Slate*.

17. R. Khan, "Steven Pinker: The Blank Slate 20+ Years Later," *Razib Khan's Unsupervised Learning* (podcast), March 17, 2023.

18. B. Sykes, *The Seven Daughters of Eve: The Science That Reveals Our Genetic Ancestry* (New York: W. W. Norton and Company, 2001).

19. D. A. Kennet, A. Timpson, D. J. Balding, and M. G. Thomas, "The Rise and Fall of BritainsDNA: A Tale of Misleading Claims, Media Manipulation and Threats to Academic Freedom," *Genealogy* 2 (2018): 47.

20. C. D. Royal, J. Novembre, S. M. Fullerton, D. B. Goldstein, J. C. Long, M. J. Bamshad, and A. G. Clark, "Inferring Genetic Ancestry: Opportunities, Challenges, and Implications," *American Journal of Human Genetics* 86, no. 5 (2020): 661–673.

21. S. Lyall, "Tracing Your Family Tree to Cheddar Man's Mum," *New York Times*, March 24, 1997.

22. S. Brace, Y. Diekmann, T. J. Booth, L. van Dorp, Z. Faltyskova, N. Rohland, S. Mallick, et al., "Ancient Genomes Indicate Population Replacement in Early Neolithic Britain," *Nature Ecology and Evolution* 3, no. 5 (2019): 765–771.

23. S. Morris, "He's One of Us: Modern Neighbours Welcome Cheddar Man," *Guardian*, February 9, 2018.

24. E. K. F. Chan, A. Timmermann, B. F. Baldi, A. E. Moore, R. J. Lyons, S.-S. Lee, A. M. F. Kalsbeek, et al., "Human Origins in a Southern African Palaeo-Wetland and First Migrations," *Nature* 575 (2019): 185–189.

25. I. Olalde, S. Mallick, N. Patterson, N. Rohland, V. Villalba-Mouco, K. Dulias, C. J. Edwards, et al., "The Genomic History of the Iberian Peninsula over the Last 8,000 Years," *Science* 363 (2019): 1230–1234.

26. A. Nelson, *The Social Life of DNA: Race, Reparations, and Reconciliation after the Genome* (Boston: Beacon Press, 2016).

27. A. Nordgren, "Genetics and Identity," *Community Genetics* 11, no. 5 (2008): 252–266.

28. M. I. Chukhryaeva, I. O. Ivanov, S. A. Frolova, S. M. Koshel, O. M. Utevska, R. A. Skhalyakho, et al., "The Haplomatch Program for Comparing Y-Chromosome STR-Haplotypes and Its Application to the Analysis of the Origins of Don Cossacks," *Russian Journal of Genetics* 52 (2016): 521–529.

29. B. Resnick, "The Limits of Ancestry DNA Tests, Explained," *Vox*, May 23, 2019.

30. K. Kampourakis, *Ancestry Reimagined: Dismantling the Myth of Genetic Ethnicities* (Oxford: Oxford University Press, 2023).

31. A. Harmon, "Why White Supremacists Are Chugging Milk (and Why Geneticists Are Alarmed)," *New York Times*, October 17, 2018.

32. W. D. Roth, "The Multiple Dimensions of Race," *Ethnic and Racial Studies* 39, no. 8 (2016): 1310–1338.

33. W. D. Roth and B. Ivemark, "Genetic Options: The Impact of Genetic Ancestry Testing on Consumers' Racial and Ethnic Identities," *American Journal of Sociology* 124, no. 1 (2018): 150–184.

34. A. Panofsky and J. Donovan, "Genetic Ancestry Testing among White Nationalists: From Identity Repair to Citizen Science," *Social Studies of Science* 49, no. 5 (2019): 653–681.

35. H. Wolinsky, "Ancient DNA and Contemporary Politics," *EMBO Reports* 20, no. 12 (2019): e49507.

36. M. Feldman, D. M. Master, R. A. Bianco, M. Burri, P. W. Stockhammer, A. Mittnik, A. J. Aja, et al., "Ancient DNA Sheds Light on the Genetic Origins of Early Iron Age Philistine," *Science Advances* 5 (2019): eaax0061.

37. H. Szeto, "Netanyahu Uses DNA Claim to Deny Palestinian Right to Homeland," *Middle East Monitor*, July 8, 2019.

38. L. Agranat-Tamir, S. Waldman, M. A. S. Martin, D. Gokhman, N. Mishol, T. Eshel, O. Cheronet, et al., "The Genomic History of the Bronze Age Southern Levant," *Cell* 181 (2020): P1146–1157.E11.

39. F. M. Riera, *Lluites antixuetes en el segle XVIII* (Barcelona: Editorial Moll, 1973).

40. J. Sharon, "Chuetas of Majorca Recognized as Jewish," *Jerusalem Post*, July 12, 2011; C. Liphshiz, "In Mallorca, Descendants of Persecuted Crypto-Jews Now Run the Community," *Times of Israel*, April 3, 2019.

41. J. F. Ferragut, J. A. Castro, C. Ramon, and A. Picornell, "Chueta Population: Diversity and Forensic Parameters of a Small, Isolated and Endogamic Population," *Forensic Science International* 7, no. 1 (2019): P496–P497; J. F. Ferragut, C. Ramon, J. A. Castro, A. Amorim, L. Alvarez, and A. Picornell, "Middle Eastern Genetic Legacy in the Paternal and Maternal Gene Pools of Chuetas," *Scientific Reports* 10 (2020): 21428.

42. T. W. Shick, "A Quantitative Analysis of Liberian Colonization from 1820 to 1843 with Special Reference to Mortality," *Journal of African History* 12, no. 1 (1971): 45–59.

43. Nelson, *The Social Life of DNA*.

44. I. Kaledzi, "Back to Roots: Why African Americans Are Flocking to Ghana," *Deutsche Welle*, January 16, 2023.

45. A. Young, "How Black America Fell out of Love with Africa," *Noēma*, March 8, 2023.

46. A. Padgen, *Peoples and Empires: A Short History of European Migration, Exploitation, and Conquest, from Greece to the Present* (New York: Random House Publishing Group, 2003).

47. K. Bryc, E. Y. Durand, J. M. Macpherson, D. Reich, and J. L. Mountain, "The Genetic Ancestry of African Americans, Latinos, and European Americans across the United States," *American Journal of Human Genetics* 96 (2015): 37–53.

48. M. D. Shriver, E. J. Parra, S. Dios, C. Bonilla, H. Norton, C. Jovel, C. Ptaff, et al., "Skin Pigmentation, Biogeographical Ancestry and Admixture Mapping," *Human Genetics* 112 (2002): 387–399.

49. C. Cillizza, "Is Barack Obama 'Black'? A Majority of Americans Say No," *Washington Post*, April 14, 2014.

50. D. R. Wilson, "Sexual Exploitation of Black Women from the Years 1619–2020," *Journal of Race, Gender and Ethnicity* 10 (2021): 122–129.

51. T. Bergin, M. Brice, N. P. Brown, D. Bryson, L. Delevingne, B. Heath, A. Januta, et al., "Slavery's Descendants Part 1; America's Family Secret," Reuters, June 27, 2023.

52. A. Hall, "Bad Blood: Hermann Goering's Niece Reveals She Had Herself Sterilised Rather than Risk Giving Birth to 'a Monster' as Relatives of Infamous Nazi Reveal How Their Family Ties Have Blighted Them," *Daily Mail*, May 9, 2016.

53. R. Morin, "An Interview with Nazi Leader Hermann Goering's Great-Niece," *Atlantic*, October 16, 2013.

54. I. Dar-Nimrod and S. J. Heine, "Genetic Essentialism: On the Deceptive Determinism of DNA," *Psychological Bulletin* 137, no. 5 (2011): 800–818.

Chapter 2

1. H. Edgeworth, *English Witnesses of the French Revolution*, ed. J. M. Thomson (London: Basil Blackwell, 1938). Memoir originally published 1815.

2. C. Lalueza-Fox, E. Gigli, C. Bini, F. Calafell, D. Luiselli, S. Pelotti, and D. Pettener, "Genetic Analysis of the Presumptive Blood from Louis XVI, King of France," *Forensic Science International* 5, no. 5 (2011): 459–463.

3. P. Charlier, I. Olalde, N. Solé, O. Ramírez, J.-P. Babelon, B. Galland, F. Calafell, et al., "Genetic Comparison of the Head of Henri IV and the Presumptive Blood from Louis XVI (Both Kings of France)," *Forensic Science International* 226, nos. 1–3 (2013): 38–40.

4. M. H. D. Larmuseau, P. Delorme, P. Germain, N. Vanderheyden, A. Gilissen, A. Van Geystelen, J.-J. Cassiman, et al., "Genetic Genealogy Reveals True Y Haplogroup of House of Bourbon Contradicting Recent Identification of the Presumed Remains of Two French Kings," *European Journal of Human Genetics* 22 (2014): 681–687.

5. E. Tucker, "Colorado Bureau of Investigation Finds DNA Scientist Manipulated Data in Hundreds of Cases over Decades," CNN, March 11, 2024.

6. S. Nurk, S. Koren, A. Rhie, M. Rautiainen, A. V. Bzikadze, A. Mikheenko, M. R. Vollger, et al., "The Complete Sequence of a Human Genome," *Science* 376, no. 6588 (2022): 44–53; A. Rhie, S. Nurk, M. Cechova, S. J. Hoyt, D. J. Taylor, N. Altemose, P. W. Hook, et al., "The Complete Sequence of a Human Y Chromosome," *Nature* 621 (2023): 344–354.

7. 1000 Genomes Project Consortium, A. Auton, L. D. Brooks, R. M. Durbin, E. P. Garrison, H. M. Kang, J. O. Korbel, et al., "A Global Reference for Human Genetic Variation," *Nature* 526 (2015): 68–74.

8. 1000 Genomes Project Consortium et al., "A Global Reference for Human Genetic Variation"; A. Biddanda, D. P. Rice, and J. Novembre, "A Variant-Centric Perspective on Geographic Patterns of Human Allele Frequency Variation," *eLife* (2020): e60107.

9. V. Tam, N. Patel, M. Turcotte, Y. Bossé, G. Paré, and D. Meyre, "Benefits and Limitations of Genome-Wide Association Studies," *Nature Review Genetics* 20 (2019): 467–484.

10. L. Fedorova, A. M. Khrunin, G. Khvorykh, J. Lim, N. Thorton, O. A. Mulyar, S. Limborska, and A. Fedorov, "Analysis of Common SNPs across Continents Reveals Major Genomic Differences between Human Populations," *Gene* 13, no. 8 (2022): 1472.

11. A. J. Jeffreys, V. Wilson, and S. L. Thein, "Individual-Specific Fingerprints of Human DNA," *Nature* 316 (1985): 76–79.

12. K. Albrecht and D. Schultheiss, "Proof of Paternity: Historical Reflections on an Andrological-Forensic Challenge," *Andrologia* 36, no. 1 (2004): 31–37.

13. K. Norrgard, "Forensics, DNA Fingerprinting, and CODIS," *Nature Education* 1, no. 1 (2008): 35.

14. I. Cobain, "Killer Breakthrough—the Day DNA Evidence First Nailed a Murderer," *Guardian*, June 7, 2016.

15. J. Curran, "Are DNA Profiles as Rare as We Think? Or Can We Trust DNA Statistics?," *Significance* 7, no. 2 (2010): 62–66.

16. K. Troyer, T. Gilboy, and B. Koeneman, "A Nine STR Locus Match between Two Apparently Unrelated Individuals Using AmpF/STR® Profiler Plus™ and Cofiler™," in *Genetic Identity Conference Proceedings*, 12th International Symposium on Human Identification, 2001, https://www.promega.ee/~/media/files/resources /conference%20proceedings/ishi%2012/poster%20abstracts/troyer.pdf.

17. B. Budowle, J. V. Planz, R. Chakraborty, T. F. Callaghan, and A. J. Eisenberg, "Clarification of Statistical Issues Related to the Operation of CODIS," in *Proceedings*

of the 17th Promega Symposium on Human Identification, 2006, https://projects.nfstc
.org/fse/pdfs/budowle.pdf.

18. Curran, "Are DNA Profiles as Rare as We Think?"

19. T. J. A. Begg, A. Schmidt, A. Kocher, M. H. D. Larmuseau, G. Runfeldt, P. A.
Maier, J. D. Wilson, et al., "Genomic Analyses of Hair from Ludwig van Beethoven,"
Current Biology 33 (2023): 1431–1447.

20. K. McKibbin, M. Shabani, and M. H. D. Larmuseau, "From Collected Stamps
to Hair Locks: Ethical and Legal Implications of Testing DNA Found on Privately
Owned Family Artifacts," *Human Genetics* 142, no. 3 (2023): 331–341.

21. A. M. Vicente, W. Ballensiefent, and J.-I. Jönsson, "How Personalized Medicine
Will Transform Healthcare by 2030: The ICPerMed Vision," *Journal of Translational
Medicine* 18 (2020): 180.

22. S. Mathur and J. Sutton, "Personalized Medicine Could Transform Healthcare,"
Biomedical Reports 7, no. 1 (2017): 3–5.

23. T. H. Saey, "Crime Solvers Embraced Genetic Genealogy," *Science News*, December 17, 2018.

24. F. Bieber, C. Brenner, and D. Lazer, "Finding Criminals through DNA of Their
Relatives," *Science* 312 (2006): 1315–1316.

25. J. Smith, "Police Are Getting DNA Data from People Who Think They Opted
Out," *Intercept*, August 18, 2023.

26. E. Niemiec and H. C. Howard, "Ethical Issues in Consumer Genome Sequencing:
Use of Consumer's Samples and Data," *Applied and Translational Genomics* 8 (2016):
22–30.

27. R. Brubaker, *Grounds for Difference* (Cambridge, MA: Harvard University Press,
2015).

28. M. H. D. Larmuseau, "Growth of Ancestry DNA Testing Risks Huge Increase in
Paternity Issues," *Nature Human Behaviour* 3 (2019): 5.

29. K. Klippenstein, "FBI Hoovering Up DNA at a Pace That Rivals China, Holds 21
Million Samples and Counting," *Intercept*, August 29, 2023.

30. K. Wang, K. Prufer, B. Krause-Kyora, A. Childebayeva, V. J. Schuenemann,
V. Coia, F. Maixner, et al., "High-Coverage Genome of the Tyrolean Iceman Reveals
Unusually High Anatolian Farmer Ancestry," *Cell Genomics* 3 (2023): 100377.

31. I. Olalde, P. Carrión, I. Mikić, N. Rohland, S. Mallick, I. Lazaridis, M. Mah, et al.,
"A Genetic History of the Balkans from Roman Frontier to Slavic Migrations," *Cell*
186, no. 26 (2023): 5705–5718.e13.

32. R. Martiniano, A. Caffell, M. Holst, K. Hunter-Mann, J. Montgomery, G. Müldner, R. L. McLaughlin, et al., "Genomic Signals of Migration and Continuity in Britain before the Anglo-Saxons," *Nature Communications* 7 (2016): 10326.

33. I. Olalde, S. Mallick, N. Patterson, N. Rohland, V. Villalba-Mouco, K. Dulias, C. J. Edwards, et al., "The Genomic History of the Iberian Peninsula over the Last 8,000 Years," *Science* 363 (2019): 1230–1234.

Chapter 3

1. M. J. Sheehan and M. W. Nachman, "Morphological and Population Genomic Evidence That Human Faces Have Evolved to Signal Individual Identity," *Nature Communications* 5 (2014): 4800.

2. J. S. Swindell, "Facial Allograft Transplantation, Personal Identity and Subjectivity," *Journal of Medical Ethics* 33, no. 89 (2007): 449–453.

3. J. G. Hall, "Twins and Twinning," *American Journal of Medical Genetics* 61, no. 3 (1996): 202–204.

4. S. B. Sabatier, M. R. Trester, and J. M. Dawson, "Measurement of the Impact of Identical Twin Voices on Automatic Speaker Recognition," *Measurement* 134 (2019): 385–389.

5. M. Teschler-Nicola, D. Fernandes, M. Händel, T. Einwögerer, U. Simon, C. Neugebauer-Maresch, S. Tangl, et al., "Ancient DNA Reveals Monozygotic Newborn Twins from the Upper Palaeolithic," *Communications Biology* 3 (2020): 650.

6. J. Diamond, *The World until Yesterday: What Can We Learn from Traditional Societies?* (London: Penguin Books, 2013).

7. R. Barquera, O. Del Castillo-Chábez, K. Nägele, P. Pérez-Ramallo, D. I. Hernández-Zaragoza, A. Szolek, A. B. Rohrlach, et al., "Ancient Genomes Reveal Insights into Ritual Life at Chicén Itzá," *Nature* 630 (2024): 912–919.

8. I. Olalde, P. Carrión, I. Mikić, N. Rohland, S. Mallick, I. Lazaridis, M. Mah, et al., "A Genetic History of the Balkans from Roman Frontier to Slavic Migrations," *Cell* 186, no. 26 (2023): 5705–5718.e13.

9. H. Jonsson, E. Magnusdottir, H. P. Eggertsson, O. A. Stefansson, G. A. Arnadottir, O. Eiriksson, F. Zink, et al., "Differences between Germline Genomes of Monozygotic Twins," *Nature Genetics* 53 (2021): 27–34.

10. M. F. Fraga, E. Ballesteros, M. F. Paz, S. Ropero, F. Setien, M. L. Ballestar, D. Heine-Suñer, et al., "Epigenetic Differences Arise during the Lifetime of Monozygotic Twins," *Proceedings of the National Academy of Sciences USA* 102, no. 30 (2005): 10604–10609.

11. H. Marcovitch, ed., *Black's Medical Dictionary*, 42nd ed. (London: A and C Black, 2010).

12. C. Pinto-Correia, *The Ovary of Eve: Egg and Sperm and Preformation* (Chicago: University of Chicago Press, 1997).

13. K. Hunter, *Duet for a Lifetime: The Story of the Original Siamese Twins* (New York: Coward-McCann, 1964); M. Bahjat, "Chang and Eng Bunker (1811–1874)," *Embryo Project Encyclopedia*, January 22, 2018.

14. J. Jacobo, "Man Wrongfully Convicted for Doppelgänger's Crime Awarded $1.1 Million," *ABC News*, December 20, 2018.

15. M. Goldberg, "Naomi Klein, Naomi Wolf and the Political Upside Down," *New York Times*, September 4, 2023.

16. N. Klein, *Doppelgänger: A Trip into the Mirror World* (London: Allen Lane, 2023).

17. F. Brunelle, "I'm Not a Look-alike!," n.d., http://www.francoisbrunelle.com/webn/e-project.html.

18. R. S. Joshi, M. Rigau, C. A. García-Prieto, M. Castro de Moura, D. Piñeyro, S. Moran, V. Davalos, et al., "Look-alike Humans Identified by Facial Recognition Algorithms Show Genetic Similarities," *Cell Reports* 40, no. 8 (2022): 111257.

19. M. J. Sheehan and M. W. Nachman, "Morphological and Population Genomic Evidence That Human Faces Have Evolved to Signal Individual Identity," *Nature Communications* 5 (2014): 4800.

20. M. Bayan, M. S. Siti, and H. Shafaatunnur, "An Overview of Uni- and Multi-Biometric Identification of Identical Twins," *IEIE Transactions on Smart Processing and Computing* 8, no. 1 (2019): 71–84.

21. P. Claes, J. Roosenboom, J. D. White, T. Swigut, D. Sero, J. Li, M. Keun Lee, et al., "Genome-Wide Mapping of Global-to-Local Genetic Effects on Human Facial Shape," *Nature Genetics* 50, no. 3 (2018): 414–423.

22. C. Attanasio, A. S. Nord, Y. Zhu, M. J. Blow, Z. Li, D. K. Biberton, H. Morrison, et al., "Fine Tuning of Craniofacial Morphology by Distant-Acting Enhancers," *Science* 342, no. 6157 (2013): 1241006.

23. Q. Li, J. Chen, M. E. Delgado, B. Bonfante, M. Fuentes-Guajardo, J. Mendoza-Revilla, J. C. Chacón-Duque, et al., "Automatic Landmarking Identifies New Loci Associated with Face Morphology and Implicates Neanderthal Introgression in Human Nasal Shape," *Communications Biology* 6 (2023): 481.

24. N. Klein, "To Know Yourself, Consider Your Doppelgänger," *New York Times*, September 13, 2023.

Chapter 4

1. A. Kjellström, "People in Transition: Life in the Mälaren Valley from an Osteological Perspective," in *Shetland and the Viking World: Papers from the Proceedings of the 17th Viking Congress 2013*, ed. V. Turner, 197–202 (Lerwick, Scotland: Shetland Amenity Trust, 2016).

2. C. Hedenstierna-Jonson, A. Kjellström, T. Zachrisson, M. Krzenwinska, V. Sobrado, N. Price, T. Günther, et al., "A Female Viking Warrior Confirmed by Genomics," *American Journal of Physical Anthropology* 164 (2017): 853–860.

3. R. Haas, J. Watson, T. Buonasera, J. Southon, J. C. Chen. S. Noe, K. Smith, et al., "Female Hunters of the Early Americas," *Science Advances* 6 (2020): eabd0310.

4. A. Anderson, S. Chilczuk, K. Nelson, R. Ruther, and C. Wall-Scheffler, "The Myth of Man the Hunter: Women's Contribution to Hunt across Ethnographic Contexts," *PLOS ONE* 18, no. 6 (2023): e0287101.

5. V. Venkataraman, J. Hoffman, R. B. Hames, D. N. E. Stibbard-Hawkes, K. Kramer, R. Kelly, K. Farquharson, et al., "Woman the Hunter? Female Foragers Sometimes Hunt, yet Gendered Divisions of Labor Are Real," preprint, BioRxiv, February 27, 2024, https://www.biorxiv.org/content/10.1101/2024.02.23.581721v1; V. Venkataraman, J. Hoffman, K. Farquharson, H. E. Davis, E. H. Hagen, R. B. Hames, B. S. Hewlett, et al., "Female Foragers Sometimes Hunt, yet Gendered Divisions of Labor Are Real: A Comment on Anderson et al. (2023) The Myth of Man the Hunter," *Evolution of Human Behavior* 45, no. 4 (2024): 106586.

6. W. Goymann, H. Nrumm, and P. M. Kappeler, "Biological Sex Is Binary, Even Though There Are a Rainbow of Sex Roles," *BioEssays* 45, no. 2 (2022): 2200173.

7. K. Jeays-Ward, M. Dandonneau, and A. Swain, "WNT4 Is Required for Proper Male and Female Sexual Development," *Developmental Biology* 276, no. 2 (2004): 431–440; P. Bernard and V. R. Harley, "WNT4 Action in Gonadal Development and Sex Determination," *International Journal of Biochemistry & Cell Biology* 39, no. 1 (2007): 31–43.

8. N. Okashita, R. Maeda, and M. Tachibana, "CDYL Reinforces Male Gonadal Sex Determination through Epigenetically Repressing WNT4 Transcription in Mice," *Proceedings of the National Academy of Sciences USA* 120, no. 20 (2022): e2221499120.

9. A. Fuentes, "Here's Why Human Sex Is Not Binary," *Scientific American*, May 1, 2023.

10. S. S. Ebenesersdóttir, M. Sandoval-Velasco, E. D. Gunnarsdóttir, A. Jagadeesan, V. B. Guðmundsdóttir, E. L. Thordardóttir, M. S. Einarsdóttir, et al., "Ancient Genomes from Iceland Reveal the Making of a Human Population," *Science* 360 (2018): 1028–1032.

11. K. Anastasiadou, M. Silva, T. Booth, L. Speidel, T. Audsley, C. Barrington, J. Buckberry, et al., "Detection of Chromosomal Aneuploidy in Ancient Genomes," *Communications Biology* 7 (2024): 14.

12. A. Rhie, S. Nurk, M. Cechova, S. J. Hoyt, D. J. Taylor, N. Altemose, P. W. Hook, et al., "The Complete Sequence of a Human Y Chromosome," *Nature* 621 (2023): 344–354.

13. J. Graves, "The 'Weird' Male Y Chromosome Has Finally Been Fully Sequenced. Can We Now Understand How It Works, and How It Evolved?," *Conversation*, August 24, 2023.

14. D. K. Griffin, "Is the Y Chromosome Disappearing?—Both Sides of the Argument," *Chromosome Research* 20, no. 1 (2012): 35–45.

15. C. Lalueza-Fox, *Inequality: A Genetic History* (Cambridge, MA: MIT Press, 2022).

16. P. A. Lee, C. P. Houk, S. F. Ahmed, and I. A. Hughes, "Consensus Statement on Management of Intersex Disorders: International Consensus Conference on Intersex," *Pediatrics* 118, no. 29 (2006): e488–500.

17. A. Fausto-Sterling, *Sexing the Body: Gender Politics and the Construction of Sexuality* (New York: Basic Books, 2000); I. A. Hughes, C. Houk, S. F. Ahmed, P. A. Lee, and LWPES1/ESPE2 Consensus Group, "Consensus Statement on Management of Intersex Disorders," *Archives Disease in Childhood* 91, no. 7 (2006): 554–563.

18. L. Sax, "How Common Is Intersex? A Response to Anne Fausto-Sterling," *Journal of Sexual Research* 39, no. 39 (2002): 174–178.

19. W. G. Reiner and P. Gearhart, "Discordant Sexual Identity in Some Genetic Males with Cloacal Assigned to Female Sex at Birth," *New England Journal of Medicine* 350 (2004): 333–341.

20. P. Paul, "As Kids, They Thought They Were Trans. They No Longer Do," *New York Times*, February 2, 2024.

21. B. Gottlieb and M. A. Trifiro, "Androgen Insensitivity Syndrome," in *GeneReviews®*, ed. M. P. Adam, J. Feldman, G. M. Mirzaa, R. A. Pagon, S. E. Wallace, and A. Amemiya (Seattle: University of Washington, 1993).

22. A. Darby, "The Case of Lydia Fairchild and Her Chimerism," *Embryo Project Encyclopedia*, 2021.

23. K. Madan, "Natural Human Chimeras: A Review," *European Journal of Medical Genetics* 63, no. 9 (2020): 103971.

24. O. Rozenblatt-Rosen, M. J. T. Stubbington, A. Regev, and S. A. Teichmann, "The Human Cell Atlas: From Vision to Reality," *Nature* 550 (2017): 451–453.

158   Notes to Chapter 4

25. R. Agate, W. Grisham, J. Wade, and S. Mann, "Neural, Not Gonadal, Origin of Brain Sex Differences in a Gynandromorphic Finch," *Proceedings of the National Academy of Sciences USA* 100, no. 8 (2002): 4873–4878.

26. J. Roughgarden, "Homosexuality and Evolution: A Critical Appraisal," in *One Human Nature: Biology, Psychology, Ethics, Politics, and Religion*, ed. M. Tibayrenc and F. J. Ayala (Cambridge, MA: Academic Press, 2016), 496–516.

27. A. C. Kinsey, W. B. Pomeroy, and C. E. Martin, *Sexual Behavior in the Human Male* (1948; repr., Bloomington: Indiana University Press, 1998).

28. T. M. Brown and E. Fee, "Alfred C. Kinsey: A Pioneer of Sex Research," *American Journal of Public Health* 93, no. 6 (2003): 896–897.

29. P. Gebhard, "Incidence of Overt Homosexuality in the United States and Western Europe," in *DHEW Publication (HMS), National Institute of Mental Health Task Force on Homosexuality: Final Report and Background Papers*, ed. J. Livingood (Washington, DC: US Department of Health, Education, and Welfare, 1972), 22–29.

30. F. Lugli, G. Di Rocco, A. Vazzana, F. Genovese, D. Pinetti, E. Chilli, M. C. Carile, et al., "Enamel Peptides Reveal the Sex of the Late Antique 'Lovers of Modena,'" *Scientific Reports* 9 (2019): 13130.

31. Roughgarden, "Homosexuality and Evolution"; J. D. Monk, E. Giglio, A. Kamath, M. R. Lambert, and C. E. Mc Donough, "An Alternative Hypothesis for the Evolution of Same-Sex Sexual Behaviour in Animals," *Nature Ecology and Evolution* 3 (2019): 1622–1631; V. Sommer and P. L. Vasey, *Homosexual Behaviour in Animals* (Cambridge: Cambridge University Press, 2006).

32. P. Chakrabarty, *Explaining Life through Evolution* (Cambridge, MA: MIT Press, 2023).

33. J. M. Gómez, A. González-Megías, and M. Verdú, "The Evolution of Same-Sex Behaviour in Mammals," *Nature Communications* 14 (2023): 5719.

34. D. Hamer, S. Hu, V. L. Magnuson, N. Hu, and A. M. Pattatucci, "A Linkage between DNA Markers on the X Chromosome and Male Sexual Orientation," *Science* 261, no. 5119 (1993): 321–327.

35. P. Conrad and S. Markens, "Constructing the 'Gay Gene' in the News: Optimism and Skepticism in the US and British Press," *Health* 5, no. 3 (2001): 373–400; G. Rice, C. Anderson, N. Risch, and G. Ebers, "Male Homosexuality: Absence of Linkage to Microsatellite Markers at Xq28," *Science* 284 (1999): 665–667.

36. R. Lewis, "Retiring the Single Gay Gene Hypothesis," *PLOS Blogs, DNA Science*, August 29, 2019.

37. A. Ganna, K. J. H. Verweij, M. G. Nivard, R. Wedow, A. S. Busch, A. Abdellaoui, S. Guo, et al., "Large-Scale GWAS Reveals Insights into the Genetic Architecture of Same-Sex Sexual Behaviour," *Science* 365, no. 64659 (2019): eaat7693.

38. T. C. Ngun, N. Ghaharamani, F. J. Sánchez, S. Bocklandt, and E. Vilain, "The Genetics of Sex Differences in Brain and Behavior," *Frontiers in Neuroendocrinology* 32, no. 2 (2010): 227–246.

39. "Gender and Health," World Health Organization, accessed January 28, 2025, https://www.who.int/health-topics/gender.

40. D. Grady, "Anatomy Does Not Determine Gender, Experts Say," *New York Times,* October 23, 2018.

41. J. Schwartz, "Is Gender Unique to Humans?," *Anthropology Magazine,* November 29, 2018.

42. O. Goldhill, "Scientific Research Shows Gender Is Not Just a Social Construct," *Quartz,* January 28, 2018.

43. B. K. Todd, R. A. Fischer, S. DiCosta, A. Roestorf, K. Harbour, P. Hardiman, and J. A. Barry, "Sex Differences in Children's Toy Preferences: A Systematic Review, Meta-Regression, and Meta-Analysis," *Infant and Child Development* 27, no. 2 (2017): e2064.

44. "Gender Incongruence and Transgender Health in the ICD," World Health Organization, accessed January 31, 2025, https://www.who.int/standards/classifications/frequently-asked-questions/gender-incongruence-and-transgender-health-in-the-icd.

45. J. Drescher, P. Cohen-Kettenis, and S. Winter, "Minding the Body: Situating Gender Identity Diagnoses in the ICD-11," *International Review Psychiatry* 24 (2012): 568–577.

46. J. G. Teisen, V. Sundaram, M. S. Filchak, L. P. Chorich, M. E. Sullivan, J. Knight, H.-G. Kim, et al., "The Use of Whole Exome Sequencing in a Cohort of Transgender Individuals to Identify Rare Genetic Variants," *Scientific Reports* 9 (2019): 20099.

47. American Psychiatric Association, "Gender Dysphoria," in *Diagnostic and Statistical Manual of Mental Disorders,* 5th ed. (Washington, DC: American Psychiatric Publishing Inc., 2013).

48. G. Heylens, G. De Cuypere, K. J. Zucker, C. Schelfaut, E. Elaut, H. V. Bossche, E. De Baere, et al., "Gender Identity Disorder in Twins: A Review of the Case Report Literature," *Journal of Sexual Medicine* 9, no. 3 (2012): 751–757.

49. G. Karamanis, M. Karalexi, R. White, T. Frisell, J. Isaksson, A. Skalkidou, and F. C. Papadopoulos, "Gender Dysphoria in Twins: A Register-Based Population Study," *Scientific Reports* 12 (2022): 13439.

50. J.-N. Zhou, M. A. Hofman, L. J. G. Gooren, and D. F. Swaab, "A Sex Difference in the Human Brain and Its Relation to Transsexuality," *Nature* 378 (1995): 68–70.

51. I. Savic and S. Arver, "Sex Dimorphism of the Brain in Male-to-Female Transsexuals," *Cerebral Cortex* 21 (2011): 2525–2533.

52. F. J. O. Boucher and T. I. Chinnah, "Gender Dysphoria: A Review Investigating the Relationship between Genetic Influences and Brain Development," *Adolescent Health Medicine and Therapeutics* 11 (2020): 89–99; J. G. Teisen, V. Sundaram, M. S. Filchak, L. P. Chorich, M. E. Sullivan, J. Knight, H.-G. Kim, et al., "The Use of Whole Exome Sequencing in a Cohort of Transgender Individuals to Identify Rare Genetic Variants," *Scientific Reports* 9 (2019): 20099.

53. T. J. C. Polderman, B. P. C. Kreukels, M. S. Irwig, L. Beach, Y.-M. Chan, E. M. Derks, I. Esteva, et al., "The Biological Contributions to Gender Identity and Gender Diversity: Bringing Data to the Table," *Behavior Genetics* 48 (2018): 95–108.

54. B. Bagemihl, *Biological Exuberance: Animal Homosexuality and Natural Diversity* (New York: St. Martin's Press, 2000).

55. D. Gonçalves, R. F. Oliveira, K. Körner, and I. Schlupp, "Intersexual Copying by Sneaker Males of the Peacock Blenny," *Animal Behaviour* 65, no. 2 (2003): 355–361.

56. D. Niven, "Male-Male Nesting Behavior in Hooded Warblers," *Wilson Bulletin* 105 (1993): 190–193.

57. R. Brubaker, *Trans: Gender and Race in an Age of Unsettled Identities* (Princeton, NJ: Princeton University Press, 2016).

58. D. St. Félix, "The Rachel Divide Review: A Disturbing Portrait of Dolezal's Racial Fraudulence," *New Yorker*, April 26, 2018.

59. V. Meade-Kelly, "Male or Female? It's Not Always So Simple," *UCLA Magazine*, August 20, 2015.

60. C. Ainsworth, "Sex Redefined: The Idea of 2 Sexes Is Overly Simplistic," *Scientific American*, October 22, 2018; A. Fuentes, "Here's Why Human Sex Is Not Binary," *Scientific American*, May 1, 2023.

Chapter 5

1. N. Dorn, "The 400th Anniversary of the Mayflower Compact," *In Custodia Legis* (Library of Congress blog), November 25, 2020, https://blogs.loc.gov/law/2020/11/the-400th-anniversary-of-the-mayflower-compact/.

2. D. P. McAdams, "The Psychology of Life Stories," *Review of General Psychology* 5 (2001): 100–122.

3. S. M. Moore, D. Rosenthal, and R. Robinson, *The Psychology of Family History: Exploring Our Genealogy* (Milton, UK: Taylor and Francis Group, 2020).

4. S. M. Moore, "Family History Research and Distressing Emotions," *Genealogy* 7, no. 2 (2023): 26.

5. F. Pessoa, *Libro del Desasosiego* (Barcelona: Sociedad Unipersonal, 2003), 208.

6. Haudenosaunee Confederacy, n.d., https://www.haudenosauneeconfederacy.com/.

7. B. Alberts, A. Johnson, J. Lewis, M. Raff, K. Roberts, and P. Walter, *Molecular Biology of the Cell*, 4th ed. (New York: Garland Sciences, 2002).

8. G. Coop, "Where Did Your Genetic Ancestors Come From?," *Coop Lab* (blog), December 19, 2017, https://gcbias.org/2017/12/19/1628/.

9. K. P. Donnelly, "The Probability That Related Individuals Share Some Section of Genome Identical by Descent," *Theoretical Population Biology* 23, no. 1 (1983): 34–63.

10. M. H. D Larmuseau, "Mommy's Baby, Daddy's Maybe: Misattributed Paternity in a Nationwide Blood Group Database," *Journal of Internal Medicine* 291, no. 1 (2022): 2–4.

11. M. H. D Larmuseau, K. Matthijs, and T. Wenseleers, "Cuckolded Fathers Rare in Human Populations," *Trends in Ecology and Evolution* 31, no. 5 (2016): 327–329.

12. B. A. Scelza, S. P. Prall, T. Blumenfield, A. N. Crittenden, M. Gurven, M. Kline, J. Koster, et al., "Patterns of Paternal Investment Predict Cross-Cultural Variation in Jealous Response," *Nature Human Behaviour* 4 (2020): 20–26.

13. Larmuseau, Matthijs, and Wenseleers, "Cuckolded Fathers Rare in Human Populations."

14. M. Peppard, "Begotten or Made? Adopted Sons in Roman Society and Imperial Ideology," in *The Son of God in the Roman World: Divine Sonship in Its Social and Political Context* (Oxford: Oxford Academic, 2011).

15. Cicero, *Pro Archia. Post Reditum in Senatu. Post Reditum ad Quirites. De Domo Sua. De Haruspicum Responsis. Pro Plancio*, trans. N. H. Watts (Cambridge, MA: Harvard University Press, 1923), 175.

16. Coop, "Where Did Your Genetic Ancestors Come From?"

17. B. Derrida, S. C. Manrubia, and D. H. Zanette, "On the Genealogy of a Population of Biparental Individuals," *Journal of Theoretical Biology* 203 (2000): 303–315.

18. G. Clark, "The Inheritance of Social Status: England, 1660 to 2022," *Proceedings of the National Academy of Sciences USA* 120, no. 27 (2023): e2300926120.

19. M. R. Robinson, A. Kleinman, M. Graff, A. A. E. Vinkhuyzen, D. Couper, M. B. Miller, W. J. Peyrot, et al., "Genetic Evidence of Assortative Mating in Humans," *Nature Human Behavior* 1 (2017): 0016; D. Conley, T. Laidley, D. W. Belsky, J. M. Fletcher, J. D. Boardman, and B. W. Domingue, "Assortative Mating and Differential Fertility by Phenotype and Genotype across the 20th Century," *Proceedings of the National Academy of Sciences USA* 113 (2016): 6647–6652.

20. H. F. Sunde, N. H. Eftedal, R. Cheesman, E. C. Corfield, T. H. Kleppesto, A. C. Seierstad, E. Ystrom, et al., "Genetic Similarity between Relatives Provides Evidence

on the Presence and History of Assortative Mating," preprint, BioRxiv, June 29, 2023, https://www.biorxiv.org/content/10.1101/2023.06.27.546663v1.

21. L. Anderson-Trocmé, D. Nelson, S. Zabad, A. Diaz-Papkovich, I. Kryukov, N. Baya, M. Touvier, et al., "On the Genes, Genealogies, and Geographies of Quebec," *Science* 380, no. 6647 (2023): 849–855.

22. M. Ferrando-Bernal, C. Morcillo-Suarez, T. de-Dios, P. Gelabert, S. Civit, A. Díaz-Carvajal, I. Ollich-Castanyer, et al., "Mapping Co-Ancestry Connections between the Genome of a Medieval Individual and Modern European," *Scientific Reports* 10, no. 1 (2020): 6843.

23. B. Ariano, V. Mattiangeli, E. M. Breslin, E. W. Parkinson, T. R. McLaughlin, J. E. Thompson, R. K. Power, et al., "Ancient Maltese Genomes and the Genetic Geography of Neolithic Europe," *Current Biology* 32, no. 12 (2022): 2668–2680.e6.

24. H. Ringbauer, Y. Huan, A. Akbari, S. Mallick, N. Patterson, and D. Reich, "AncIBD-Screening for Identity by Descent Segments in Human Ancient DNA," preprint, BioRxiv, March 9, 2023, https://www.biorxiv.org/content/10.1101/2023 .03.08.531671v1.

25. E. Harney, S. Micheletti, K. S. Bruwelheide, W. A. Freyman, K. Bryc, A. Akbari, E. Jewett, et al., "The Genetic Legacy of African Americans from Catoctin Furnace," *Science* 381 (2023): eade4995.

26. A. Nelson, *The Social Life of DNA: Race, Reparations, and Reconciliation after the Genome* (Boston: Beacon Press, 2016).

27. Harney et al., "The Genetic Legacy of African Americans from Catoctin Furnace."

28. R. Curry, "Ethics and the Study of Historic DNA of African Americans Buried at the Catoctin Furnace," *23andMe* (blog), August 3, 2023.

29. Nelson, *The Social Life of DNA*.

30. New England Historic Genealogical Society, "Mayflower Families Fifth Generation Descendants, 1700–1800 Is Now Complete," American Ancestors, May 14, 2018, https://dbnews.americanancestors.org/2018/05/14/mayflower-families-fifth-gen eration-descendants-1700-1880-is-now-complete/.

31. A. Curry, "DNA from Enslaved 19th Century Maryland Girl Traced Forward to Living Relatives," *Science*, https://doi.org/10.1126/science.zcn31q4.

Chapter 6

1. J. Bradbury, *The Capetians; Kings of France 987–1328* (London: Bloomsbury Publishing 2007).

2. R. Bartlett, *Blood Royal: Dynastic Politics in Medieval Europe* (New York: Cambridge University Press, 2020).

3. F. C. Ceballos and G. Alvarez, "Royal Dynasties as Human Inbreeding Laboratories: The Habsburgs," *Heredity* 111, no. 2 (2013): 114–121.

4. Bartlett, *Blood Royal*.

5. N. Cummins, "Lifespans of the European Elite, 800–1800," *Journal of Economic History* 77, no. 2 (2017): 406–439.

6. B. Derrida, S. C. Manrubia, and D. H. Zanette, "Statistical Properties of Genealogical Trees," *Physical Review Letters* 82 (1999): 1987–1990.

7. R. H. Bixler, "Sibling Incest in the Royal Families of Egypt, Peru and Hawaii," *Journal of Sex Research* 18 (1982): 264–281.

8. R. Middleton, "Brother-Sister and Father-Daughter Marriage in Ancient Egypt," *American Sociological Review* 27 (1962): 603–611.

9. N. Reeves, *Akhenaten: Egypt's False Prophet* (London: Thames and Hudson, 2005).

10. S. L. Ager, "The Power of Excess: Royal Incest and the Ptolemaic Dynasty," *Anthropologica* 48 (2006): 165–186.

11. S. L. Ager, "Familiarity Breeds: Incest and the Ptolemaic Dynasty," *Journal of Hellenic Studies* 125 (2005): 1–34.

12. J. Tyldesley, *Cleopatra: Last Queen of Egypt* (New York: Basic Books, 2008).

13. A. Michalopoulos, G. Tzelepis, and S. Geroulanos, "Morbid Obesity and Hypersomnolence in Several Members of an Ancient Royal Family," *Thorax* 58 (2003): 281–282.

14. Ceballos and Alvarez, "Royal Dynasties as Human Inbreeding Laboratories."

15. D. Kleinman-Ruiz, M. Lucena-Perez, B. Villanueva, J. Fernández, A.P. Saveljev, M. Ratkiewicz, K. Schmidt, et al., "Purging of Deleterious Burden in the Endangered Iberian Lynx," *Proceedings of the National Academy of Sciences USA* 119, no. 11 (2022): e2110614119.

16. G. Alvarez, F. C. Ceballos, and C. Quinteiro, "The Role of Inbreeding in the Extinction of a European Royal Dynasty," *PLOS ONE* 4 (2009): e5174.

17. G. Parker, *Emperor: A New Life of Charles V* (New Haven, CT: Yale University Press, 2016).

18. R. Vilas, F. C. Ceballos, L. Al-Soufi, R. González-García, C. Moreno, M. Moreno, L. Villanueva, et al., "Is the 'Habsburg Jaw' Related to Inbreeding?," *American Journal of Human Biology* 46, no. 7–8 (2019): 553–561.

19. G. Alvarez, R. Vilas, F. C. Ceballos, H. Carvalhal, and T. J. Peters, "Inbreeding in the Last Ruling Dynasty of Portugal: The House of Braganza," *American Journal of Human Biology* 31, no. 2 (2018): e23210.

20. T. J. Peters and C. Willis, "Mental Health Issues of Maria I of Portugal and Her Sisters: The Contributions of the Willis Family to the Development of Psychiatry," *History of Psychiatry* 24, no. 3 (2013): 292–307.

21. R. Stevens, "The History of Haemophilia in the Royal Families of Europe," *British Journal of Haematology* 105 (1999): 25–32.

22. W. Shakespeare, *The Tragedy of King Richard III* (1597; Project Gutenberg, 1998), e-book 1503, www.Gutenberg.org/files/1503.

23. T. King, G. Gonzalez Fortes, P. Balaresque, M. G. Thomas, D. Balding, P. Maisano Delser, R. Neumann, et al., "Identification of the Remains of King Richard III," *Nature Communications* 5 (2014): 5631.

24. B. Jones, "Richard III Was Blue-Eyed, Blond, but Should He Have Been King? DNA Puzzle," CNN, December 5, 2014.

25. S. Husband, "Pedigree Chums: A History of Aristocratic Incest," *Rake*, September 2015.

26. F. Morton, *The Rothschilds: A Family Portrait* (New York: Diversion Books, 2014).

27. R. Conniff, "Go Ahead, Kiss Your Cousin," *Discovery Magazine*, August 1, 2003.

28. L. M. Cassidy, R. O. Maoldúin, T. Kador, A. Lynch, C. Jones, P. C. Woodman, E. Murphy, et al., "A Dynastic Elite Monumental Neolithic Society," *Nature* 582 (2020): 384–388.

29. H. Keen, *The Science of Game of Thrones: From the Genetics of Royal Incest to the Chemistry of Death by Molten Gold—Sifting Fact from Fantasy in the Seven Kingdoms* (Boston: Little, Brown and Company, 2016).

30. F. C. Ceballos, K. Gürün, N. E. Atinisik, H. C. Gemici, C. Karamurat, D. Koptekin, K. B. Vural, et al., "Human Inbreeding Has Decreased in Time through the Holocene," *Current Biology* 31 (2021): 3925–2934.e8.

31. J. Kaplanis, A. Gordon, T. Shor, O. Weissbrod, D. Geiger, M. Wahl, M. Gershowitz, et al., "Quantitative Analysis of Population-Scale Family Trees with Millions of Relatives," *Science* 360, no. 6385 (2018): 171–175.

32. F. C. Ceballos, P. K. Joshi, D. W. Clark, M. Ramsay, and J. F. Wilson, "Runs of Homozygosity: Windows into Population History and Trait Architecture," *Nature Reviews Genetics* 19 (2018): 220–234.

33. D. W. Owsley, K. S. Bruwelheide, E. Harney, S. Mallick, N. Rohland, I. Olalde, K. G. Barca, et al. "Historical and Archaeogenomic Identification of High-Status

Englishmen at Jamestown, Virginia," *Antiquity* 98, no. 400 (2024), https://doi.org/10 .15184/aqy.2024.75.

Chapter 7

1. Quoted in R. J. Richards, "The Beautiful Skulls of Schiller and the Georgian Girl: Quantitative and Aesthetic Scaling of the Races, 1770–1850," in *Johann Friedrich Blumenbach: Race and Natural History, 1750–1850*, ed. N. Rupke and G. Lauer (London: Taylor and Francis Group, 2019), 142.

2. N. Rupke and G. Lauer, "Introduction: A Brief History of Blumenbach Representation," in *Johann Friedrich Blumenbach: Race and Natural History, 1750–1850*, ed. N. Rupke and G. Lauer (London: Taylor and Francis Group, 2019), 3–15.

3. A. B. Popejoy, "Too Many Scientists Still Say Caucasian," *Nature*, August 26, 2021.

4. C. Lalueza, *Razas, racismo y diversidad: La ciencia, un arma contra el racismo* (Alzira, Spain: Bromera Ed-Universitat de Valencia, 2001); J. Marks, "Race, Past, Present and Future," in *Revisiting Race in a Genomic Age*, ed. B. Koenig, S. Lee, and S. Richardson (New Brunswick, NJ: Rutgers University Press, 2008), 21–38.

5. B. Kenyon-Flatt, "How Scientific Taxonomy Constructed the Myth of Race," *Anthropology Magazine*, March 19, 2021.

6. W. Jankowsky, "Konstitution, Körperbau und Rasse in ihrer gegenseitigen Beziehung und Abgrenzung," *Anatomischer Anzeiger* 70 (1930): 470–515; translation by author.

7. C. Darwin, *The Descent of Man and Selection in Relation to Sex* (London: J. Murray, 1871).

8. S. Rose, "Darwin, Race and Gender," *EMBO Reports* 10, no. 4 (2009): 297–298.

9. J. Marks, "Ten Facts about Human Variation," in *Human Evolutionary Biology*, ed. M. Muelenbein (Cambridge: Cambridge University Press, 2010), 265–276.

10. J. L. Graves, *The Race Myth: Why We Pretend Race Exists in America* (New York: Dutton, 2004).

11. R. Pérez Ortega, "Human Geneticists Curb Use of the Term 'Race' in Their Papers," *Science*, December 2, 2021.

12. T. H. Eriksen, *Small Places, Large Issues: An Introduction to Social and Cultural Anthropology*, 3rd ed. (London: Pluto Press, 1995).

13. E. E. Evans-Pritchard, *The Nuer* (Oxford: Clarendon, 1940).

14. P. P. Howell, *A Manual of Nuer Law* (Oxford: Oxford University Press, 1954).

15. D. Sneath, "Tribe," in *The Open Encyclopedia of Anthropology*, ed. F. Stein, September 1, 2016, https://doi.org/10.29164/16tribe.

16. Eriksen, *Small Places, Large Issues*.

17. A. Saville, "Anatomizing an Archaeological Project—Hazleton Revisited," *Translations of Bristol and Gloucestershire Archaeological Society* 128 (2010): 9–27.

18. C. Fowler, I. Olalde, V. Cummings, I. Armit, L. Büster, S. Cuthbert, N. Rohland, et al., "A High-Resolution Picture of Kinship Practices in an Early Neolithic Tomb," *Nature* 601 (2022): 584–587.

19. L. Stone and D. E. King, *Kinship and Gender: An Introduction* (Milton Park, UK: Routledge, 2018).

20. M. Rivollat, A. B. Rohlach, H. Ringbauer, A. Childebayeva, F. Mendisco, R. Barquera, A. Szolek, et al., "Extensive Pedigrees Reveal the Social Organization of a Neolithic Community," *Nature* 620 (2023): 600–606.

21. S. Mallick, H. Li, M. Lipson, I. Mathieson, M. Gymrek, F. Racimo, M. Zhao, et al., "The Simons Genome Diversity Project: 300 Genomes from 142 Diverse Populations," *Nature* 538 (2016): 201–206.

22. R. Lewontin, "The Apportionment of Human Diversity," in *Evolutionary Biology*, ed. T. Dobzhansky, M. K. Hecht, and W. C. Steer (New York: Springer, 1972), 6:381–398.

23. A. W. F. Edwards, "Human Genetic Diversity: Lewontin's Fallacy," *Bioessays* 25, no. 8 (2003): 798–801.

24. J. Kitchens and G. Coop, "Visualizing Human Genetic Diversity," *James Kitchens* (blog), May 16, 2023, https://james-kitchens.com/blog/visualizing-human-genetic-diversity.

25. L. L. Cavalli-Sforza and A. W. F. Edwards, "Analysis of Human Evolution," *Genetics Today* 3 (1964): 923–933.

26. J. Marks, "We're Going to Tell Those People Who They Really Are," in *Relative Values: Reconfiguring Kinship Studies*, ed. S. Franklin and S. McKinnon (Chapel Hill, NC: Duke University Press, 2002), 355–383.

27. Y. J. J. Byeon, R. Islamaj, L. Yeganova, W. J. Wilbur, Z. Lu, L. C. Brody, and V. L. Bonham, "Evolving Use of Ancestry, Ethnicity, and Race in Genetics Research—a Survey Spanning Seven Decades," *American Journal of Human Genetics* 108, no. 12 (2021): 2215–2223.

28. Marks, "Race, Past, Present and Future."

29. R. Brubaker and F. Cooper, "Beyond 'Identity,'" *Theory and Society* 29, no. 1 (2000): 1–47.

30. D. Serre and S. Pääbo, "Evidence for Gradients of Human Genetic Diversity within and among Continents," *Genome Research* 14, no. 9 (2004): 1679–1685.

31. Race, Ethnicity, and Genetics Working Group, "The Use of Racial, Ethnic, and Ancestral Categories in Human Genetics Research," *American Journal of Human Genetics* 77, no. 4 (2005): P519–P532.

32. L. B. Jorde and M. J. Bamshad, "Genetic Ancestry Testing: What Is It and Why Is It Important?," *Journal of the American Medical Association* 323, no. 11 (2020): 1089–1090.

33. D. Reich, *Who We Are and How We Got Here: Ancient DNA and the New Science of the Human Past* (New York: Pantheon Books, 2018).

34. C. D. Royal, J. Novembre, S. M. Fullerton, D. B. Goldstein, J. C. Long, M. J. Bamshad, and A. G. Clark, "Inferring Genetic Ancestry: Opportunities, Challenges, and Implications," *American Journal of Human Genetics* 86, no. 5 (2020): 661–673.

35. I. Mathieson and A. Scally, "What Is Ancestry?," *PLOS Genetics* 16 (2020): e1008624.

36. Mathieson and Scally, "What Is Ancestry?"

37. G. Coop, "Genetic Similarity versus Genetic Ancestry Groups as Sample Descriptors in Human Genetics," preprint version 2, arXiv, last revised January 7, 2023, https://arxiv.org/abs/2207.11595v2.

38. M. Thomas, "To Claim Someone Has 'Viking Ancestors' Is No Better than Astrology," *Guardian*, February 25, 2013.

39. Reich, *Who We Are and How We Got Here*.

40. S. Geroulanos, *The Invention of Prehistory: Empire, Violence and Our Obsession with Human Origins* (New York: Liveright Publishing Corporation, 2024).

Chapter 8

1. G. Kolata, *Clone: The Road to Dolly and the Path Ahead* (London: Penguin Books, 1997).

2. H. T. Greely, "Human Reproductive Cloning: The Curious Incident of the Dog in the Night-Time," *Stat News*, February 21, 2020.

3. J. Brogan, "The Real Reasons You Shouldn't Clone Your Dog," *Smithsonian Magazine*, March 22, 2018.

4. B. Streisand, "Barbra Streisand Explains: Why I Cloned My Dog," *New York Times*, March 2, 2018.

5. J. Cohen, "As Creator of 'CRISPR Babies' Nears Release from Prison, Where Does Embryo Editing Stand?," *Science*, March 21, 2022.

6. G. Alanis-Lobato, J. Zohren, A. McCarthy, N. M. E. Fogarty, N. Kubikova, E. Hardman, M. Greco, et al., "Frequent Loss of Heterozygosity in CRISPR-Cas9-Edited Early Human Embryos," *Proceedings of the National Academy of Sciences USA* 118, no. 22 (2021): e2004832117.

7. H. Eiberg, J. Troelsen, M. Nielsen, A. Mikkelsen, J. Mengel-From, K. W. Kjaer, and L. Hansen, "Blue Eye Color in Humans May Be Caused by a Perfectly Associated Founder Mutation in a Regulatory Element Located within the HERC2 Gene Inhibiting OCA2 Expression," *Human Genetics* 123, no. 2 (2008): 177–187.

8. I. Olalde, M. E. Allentoft, F. Sánchez-Quinto, G. Santpere, C. W. K. Chiang, M. DeGiorgio, J. Prado-Martínez, et al., "Derived Immune and Ancestral Pigmentation Alleles in a 7,000-Year-Old Mesolithic European," *Nature* 507 (2014): 225–228.

9. M. de Montaigne, *Essays of Michel de Montaigne*, ed. William Carew Hazlitt, trans. Charles Cotton (1877; Project Gutenberg, 2004), bk. 3, chap. 2, www.gutenberg.org /ebooks/3600.

10. G. Coop, "Genetic Similarity and Genetic Ancestry Groups," preprint version 1, arXiv, July 23, 2022, https://arxiv.org/abs/2207.11595v1.

11. P. Wade, C. López-Beltrán, and R. Ventura Santos, "Genomic Research, Publics and Experts in Latin America: Nation, Race and Body," *Social Studies of Science* 45, no. 6 (2015): 775–796.

12. M. Griffiths, "Identity," *Oxford Bibliographies*, June 29, 2015, https://www.oxford bibliographies.com/display/document/obo-9780199766567/obo-9780199766567 -0128.xml.

13. B. Anderson, *Imagined Communities: Reflections on the Origin and Spread of Nationalism* (London: Verso, 1991).

14 L. Bos, C. Shemer, N. Corbu, M. Hameleers, I. Andreadis, A. Schulz, D. Schmuck, et al., "The Effects of Populism as a Social Identity Frame on Persuasion and Mobilization: Evidence from a 15-Country Experiment," *European Journal of Political Research* 59, no. 1 (2020): 3–24.

Index

Publisher contact:
The MIT Press
Massachusetts Institute of Technology
77 Massachusetts Avenue, Cambridge, MA 02139
mitpress.mit.edu

EU Authorised Representative:
Easy Access System Europe, Mustamäe tee 50,
10621 Tallinn, Estonia
gpsr.requests@easproject.com

Printed by Integrated Books International,
United States of America